AToMech1-2023 Supplement

AToMech1-2023
International Conference on Advanced Topics in Mechanics of
Materials, Structures and Construction
March 12-14, 2023, College of Engineering, Prince Mohammad
Bin Fahd University, Al Khobar, Saudi Arabia

Editors
Erasmo Carrera
Faramarz Djavanroodi
Muhammad Asad

Peer review statement

All papers published in this volume of "Materials Research Proceedings" have been peer reviewed. The process of peer review was initiated and overseen by the above proceedings editors. All reviews were conducted by expert referees in accordance to Materials Research Forum LLC high standards.

Published under License by **Materials Research Forum LLC**
Millersville, PA 17551, USA

Published as part of the proceedings series
Materials Research Proceedings
Volume 36 (2023)

ISSN 2474-3941 (Print)
ISSN 2474-395X (Online)

ISBN 978-1-64490-278-3 (Print)
ISBN 978-1-64490-279-0 (eBook)

Distributed worldwide by

Materials Research Forum LLC
105 Springdale Lane
Millersville, PA 17551
USA
https://www.mrforum.com

Manufactured in the United State of America
10 9 8 7 6 5 4 3 2 1

Table of Contents

Preface

On behalf of the conference committee, I would like to thank all the participants in the International Conference on Advanced Topics in Mechanics of Materials, Structures and Construction (AToMech1-2023) which had been held at Prince Mohammad Bin Fahd University (PMU) Khobar, Kingdom of Saudi Arabia on March, 2023. I was honored to serve as Chairman of this important event. We are all aware that research plays an important role in its contribution to the knowledge, development and shaping of the future directions.

While globalization appeared to have significant impacts for the worldwide society, AToMech1 provided a platform to promote sustainable development, international networking, for researchers, practitioners and educators all over the world. The conference offered a truly comprehensive view while inspiring the attendees to come up with solid recommendations to tackle hot-topic challenges. Finally, I thank the keynote speakers; presenters and authors for contribution.

Dr. Faisal Yousif Al Anezi
Conference Chair

Committees

Honory Chair

Issa H. Alansari, President, Prince Mohammed Bin Fahd University

Conference Chair

Dr. Faisal Yousif Al Anezi, Vice President, Prince Mohammed Bin Fahd University

Conference Co – Chair

Professor Erasmo Carrera, Prince Mohammed Bin Fahd University

Technical Program Committee

Dr. J. N. Reddy, Texas A&M University, USA
Dr. Jamal Nayfeh, Prince Mohammad Bin Fahd University, KSA
Dr. Erasmo Carrera, Politecnico di Torino, Italy
Dr. Adel M. Hanna, University of Concordia, Canada
Dr. Chandra Sekhar Matli, National Institute of Technology, India
Dr. Siti Noor Linda bt Taib, Universiti Malaysia Sarawak, Malaysia
Dr. Faramarz Djavanroodi, Prince Mohammad Bin Fahd University, KSA
Dr. Weiqiu Chen, Zhejiang University, PR China
Dr. Muhammad Asad, Prince Mohammad Bin Fahd University, KSA
Dr. Volodymyr Zozulya, Kharkiv State Technical University, Ukraine
Dr. Esam Jassim, Prince Mohammad Bin Fahd University, KSA
Dr. Muhammad Azhar Khan, Prince Mohammad Bin Fahd University, KSA
Dr. Mabrouk Touahmia, University of Hail, KSA
Dr. Muhammad Kalimur Rahman, King Fahd University of Petroleum and Minerals, KSA
Dr. Mustafa Aytekin, University of Bahrain, Bahrain
Dr. Xin Ren, Nanjing Tech University, PR China
Dr. Jose Luis Mantari Laureano, Universidad De Ingeniera Y Tecnologia, Peru
Dr. Mushtaq Khan, Prince Mohammad Bin Fahd University, KSA
Dr. Adegbola Akinola, Obafemi Awolowo University, Nigeria
Dr. Bouchaib Zazoum, Prince Mohammad Bin Fahd University, KSA
Dr. O. P. Layeni, Obafemi Awolowo University, Nigeria
Dr. Andi Asiz, Prince Mohammad Bin Fahd University, KSA
Dr. Xiangyang Xu, Soochow University, PR China
Dr. Li xiangyu, Southwest Jiaotong University, PR China
Dr. Alfonso Pagani, Politecnico di Torino, Italy
Dr. Mário Rui Arruda, Instituto Tecnico Lisboa, Portugal
Dr. Tahar Ayadat, Prince Mohammad Bin Fahd University, KSA
Dr. Xin Fang, National University of Defence Technology, PR China
Dr. Maiaru Marianna, University of Massachusetts, USA
Dr. Feroz Shaik, Prince Mohammad Bin Fahd University, KSA
Dr. Mouhammad El Hassan, Prince Mohammad Bin Fahd University, KSA
Dr. Yousef Alshammari, Prince Mohammad Bin Fahd University, KSA
Dr. Malek Brahimi, New York City College of Technology, USA,
Dr. Syed Sohail Akhtar, King Fahd University of Petroleum and Minerals, KSA

Technical Program Chairs

1. Session Title: Novel Metamaterial Structures

Organizer: Dr. Xin Ren, xin.ren@njtech.edu.cn Nanjing Tech University, PR China

2. Session Title: Structural analysis with machine learning.

Organizer:Dr. Jose L. Mantari Laureano, josemantari@gmail.com Universidad De Ingenieriay Technologia, Perù

3. Session Title: Tensor Application and Constitutive Equation Modelling in Continuum Mechanics and Engineering Design

Organizers: Dr. Adegbola Akinola, aakinola@oauife.edu.ng . Dr. O. P. Layeni, olayeni@oauife.edu.ng Obafemi Awolowo University, Nigeria

4. Session Title: Intelligent monitoring of engineering structure

Organizer: Dr. Xiangyang Xu, xuxiangyang_ca@163.com Soochow University, PR, China

5. Session title: Mechanics of functional materials

Organizer: Dr. Li xiangyu, lixiangyu@swjtu.edu.cn Southwest Jiaotong University, PR China

6. Session Title: Nonlinear structure mechanics and failure analysis

Organizer: Dr. Alfonso Pagani, alfonso.pagani@polito.it Politecnico di Torino, Italy Dr. Mário Rui Arruda, mario.rui.arruda@etecnico.ulisboa.pt Instituto Tecnico Lisboa, Portugal

7. Session Title: Programmable gear-based mechanical metamaterials

Organizer: Dr. Xin Fang, xinfangdr@sina.com National University of Defence Technology, PR China

8. Session Title: Multiscale Analysis of Structures and Materials

Organizer: Dr. Maiaru Marianna, Marianna_Maiaru@uml.edu University of Massachusetts, USA

9. Session Title: Mechanics of materials and structures produced by additive manufacturing technologies

Organizer: Dr. Mohammad Asad, masad@pmu.edu.sa Prince Mohammad Bin Fahad University, KSA

10. Session Title: Numerical modelling and experimental testing of materials and structures

Organizer: Dr. Muhammad Azhar Khan, mkhan6@pmu.edu.sa Prince Mohammad Bin Fahad University, KSA

11. Session Title: Advanced Construction Materials in Geotechnical Applications, Structures and Foundation Engineering

Organizer: Dr. Tahar Ayadat, tayadat@pmu.edu.sa Prince Mohammad Bin Fahad University, KSA

12. Session Title: Future Sustainable Infrastructure and Construction Materials

Organizer: Dr. Andi Asiz, aasiz@pmu.edu.sa Prince Mohammad Bin Fahad University, KSA

13. Session Title: Fluid-Structure Interaction

Organizer: Dr. Esam Jassim, ejassim@pmu.edu.sa Prince Mohammad Bin Fahad University, KSA

14. Session Title: Modelling and Optimization of Materials Manufacturing Processes

Organizer: Dr. Kuldeep Saxena, saxena0081@gmail.com Research and Development, Lovely Professional University, Phagwara, India

AToMech1-2023 Supplement
Materials Research Proceedings 30 (2023) 1-7

Materials Research Forum LLC
https://doi.org/10.21741/9781644902790-1

Hand gesture control accuracy through increased epoch and batch size

AJAYI Oluwaseun Kayode[1,a*], AJAYI Oladipupo Omogbolahan[1,b],
MALOMO Babafemi O.[1,c], ADEYI Abiola John[2,d] and NWANKWO Benneth O.[1,e]

[1]Department of Mechanical Engineering, Obafemi Awolowo University, Ile-Ife, Nigeria

[2]Department of Mechanical Engineering, Ladoke Akintola University of Technology, Ogbomoso, Nigeria

[a]okajayi@oauife.edu.ng, [b]ajayidipo@ymail.com, [c]bobmalom@yahoo.com, [d]adeyi.abiola@yahoo.com and [e]nwankwobenneth22@gmail.com

Keywords: Tensor Flow, Facial Recognition, Batch Number, Epoch

Abstract. The conventional method of mechanism control which requires physical contact is being replaced with remote control, especially with the advent of the Internet of things (IoT). Facial recognition is used to identify and authenticate faces from images or videos. It has many applications, including granting or restricting access to a facility, or secured areas, or preventing unauthorized users, including selective access. The accuracy of such a recognition and control system is very important and depends on how the system was modeled and adopted. In this study, facial images and gesture images were collated and modeled in a python neural network, then optimized using TensorFlow. The algorithm was compiled unto a raspberry pi for testing on a developed automatic gate. An effective method of achieving accuracy using the epoch is presented in this work. Five (0, 5, 10, 15, 20) batch number and epoch respectively were modeled to achieve accuracy for the system. The model was trained with five gestures; fist, palm, thumb up, thumb, and last finger. The gesture recognition accuracy achieved through the epoch was maximum at 99.97 when both batch number and epoch were set to 20. However, when no epoch was set, the accuracy was below 10.2, whereas there was no accuracy below 98% when the epoch was introduced. This depicts the importance of epoch in achieving accuracy in image recognition. It was also discovered that the higher the epoch and the batch number the greater the accuracy, but the processing would require a very high processing unit.

Introduction

The conventional method of mechanism control which requires physical contact is being replaced with remote control, especially with the advent of the Internet of things (IoT). The concept of biometrics involves capturing biological images, storing them, and later retrieving them usually as a form of registering the object. Facial recognition, a biometrics concept is used to identify and authenticate faces either from images or videos. It has many applications, including granting or restricting access to a facility, secured areas or prevention of unauthorized users including selective access [1-5]. Among the pioneer works in facial recognition was the work of Sirovich and Kirby who presented a solution to solve facial algebra problems using linear algebra, but were limited in capacity to store large data [6]. Despite the early efforts, this system gained traction early 2000s with the proliferation of mobile telephony, larger computer power, and data storage [7].

To achieve facial recognition, feature extraction is paramount, which is divided into face and facial landmark extraction [8]. The accuracy of detection can be governed using Bayesian rule-based, Gaussian mixture model and the Expected-maximization (EM) algorithms [9, 10]. The outputs are in form of statistical values or otherwise referred to as probabilistic values, which depict the accuracy of the pixel detected [10]. This work presents the effect of epoch and batch number on the detection of face and hand gestures in the control of an automatic gate mechanism. Real-time tracking of hand gestures has been worked on in recent times [11, 12], while a combination of hand gestures and head pose was adopted for a control system [13]. These two have found great applications in home security and automation [14].

Methodology

This section involves three stages; Facial recognition (data collection, data pre-processing, and model training and validation), Hand gesture recognition, and test with Automatic garage door.

Facial Recognition: This stage was achieved through; Data Collection, Data Pre-processing and Model training, and Validation.

For the data collection, data were obtained from the OpenFace database; 20 images of 5 different faces were added and classified to make 100 images processed for testing.

Data pre-processing for detection was in two phases; face detection and facial landmark detection. In face detection, the Haar cascade algorithm (OpenCV) [15] was used from the python library utilizing the Single Shot Detector (SSD) framework with the ResNet as the base network, while for face landmark detection; dlib and OpenCV were used.

Face Detection: This phase involves data collection, data pre-processing, model training and validation, and model evaluation. Deep neural network was used to train the face recognition algorithm. Images were inputted 2000 per batch which includes; anchor image, positive image, and negative images for each element of the batch.

Model training and classification: Deep neural network in python, which usually involves data entry and model training. In this procedure, triplet loss function was used for the training which has the advantage of ensuring that the anchor image is closer to the positive image (real image) than it is to the negative image. In the training, three sets of images; anchor image, positive image, and negative image constituted the inputs. In the image input, anchor and positive images (image belonging to the same person) and negative image (image belonging to another person) were inputted. 123-dimensions embedding for each face is computed and assigned weights through the network which results in larger disparity between the real images and the negative image.

$$\sum_i^N \left[||f(x_i^a) - f(x_i^p)||_2^2 - ||f(x_i^a) - f(x_i^n)||_2^2 + \alpha \right] \qquad 1$$

The system was built to work real-time using the haar cascade face detection algorithm. After detection, it takes a picture of the face, runs through face encoding, comparing with existing or registered users.

Hand gesture recognition: The Convolutionary neural network deep algorithm has been handy for multiple image processing; therefore, it was adopted for the hand gesture recognition modeling. The workflow order adopted was data collection, data pre-processing and feature recognition, and training machine learning model.

First set of data was collected for open palm, fist, index finger, index finger and thumb, thumb up, thumb down and three middle fingers by taking physical images while the second batch was gotten online form Kaggle.com. Therefore, 4000 images were collected for each gesture. The data was reset to 50 x 50 pixels for uniformity. This data was used to train the system using deep neural network in python program. Two thousand images each on open palm, closed palm (fist), index finger, index finger and thumb, thumb up, thumb down, and three middle fingers were sourced making up fourteen thousand images from where three thousand were chosen for validation purposes. To make the model more robust, the position and size of the gestures were varied for each frame. These pictures were either self-taken or sourced from kaggle.com. In the data pre-processing, images were set to 50 x 50 order for uniformity. The training has input size of 2500 nodes with 25 nodes corresponding to 5 nodes of 5 hand gestures.

AToMcoh1 2023 Supplement
Materials Research Proceedings 36 (2023) 1-7

Materials Research Forum LLC
https://doi.org/10.21741/9781644902790-1

Automatic garage door prototype: Materials: raspberry pi, camera module, python, pc and automatic gate.

A prototype automatic garage door using the materials mentioned above was assembled to evaluate the performance of facial recognition and hand gesture intelligence. The algorithm was compiled on a raspberry pi to control the opening and closing of the gate the camera module was attached to the top of the gate to detect a registered face, the identified face is then given access to control the gate through hand gestures to open, stop or close the gate. The microcontroller sends a corresponding hand gesture signal through the Infrared module to the servo motor to either open or close the gate. Therefore, the algorithm order is; face detection, face authentication, hand gesture detection, send detected hand gesture to the servo motor and servo motor carries out sent instruction/ command. This procedure is repeated for every detection.

The trained and created model's application file was then uploaded to the raspberry pi, which has a camera module for facial and gesture detection.

Results

In training the hand gesture recognition model, the following parameters were set; 2D convolution layer applying number of convolution filters to the image: three pooling layers were used to downsample images extracted by the convolutional layers thereby reducing the dimensionality of the feature map to achieve a reduction in the processing time; introduction of dense layers which performs classification on the layers in previous procedures; the use of batch size which controls the accuracy of the estimated error gradient in training neural networks: number of epochs that determines whether the algorithm will work through the entire training dataset; and the number of trained images used per gesture to train the hand gesture model.

Tensorflow was used to define the Deep Neural Network Model used to train the hand gesture recognition model. Five input arguments were used; Data to train the model, Target to train the model, Number of epochs, validation sets/data, and batch size. The variation of batch sizes 5, 10, 15 and 20 were used against epoch variations 0, 5, 10, 15, and 20. Model accuracy and time taken to train the model were recorded and presented in Figures 1, Figure 2, and 3. The response time varied between 40.89 seconds at 5 epoch and batch sizes respectively to 1892.42 when batch size and epoch were set to 20 each. However, the accuracy of the model was highest at the maximum epoch and batch sizes with the value at 99.967 as seen in table 1.

Higher epoch and batch sizes would have been used, but for the limitation of the CPU, which has the likelihood of getting an accuracy closer to 100% for the model identification of images but with higher computation time as presented in table 2 and table 3. The facial recognition system was able to detect multiple registered users at the same time (Figure 4).

Table 1: Hand gesture model for five users

	Confidence Level				
Users	1	2	3	4	5
1	99.92828	99.76700	99.59408	99.03502	99.57861
2	99.05941	99.07965	99.01235	99.48297	99.83151
3	98.68325	98.27860	99.08360	99.56101	98.64261
4	99.86299	99.00904	99.61547	99.34412	99.50295
5	98.15253	98.54012	98.24575	98.87617	98.59599

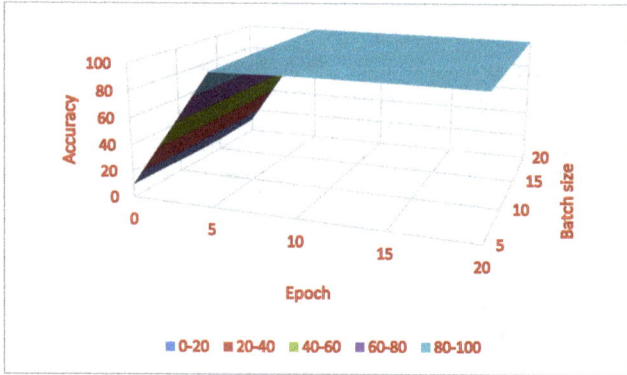

Figure 1: Accuracy level with varying batch size and epoch

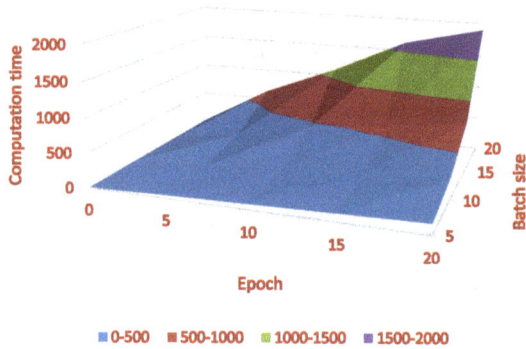

Figure 2: Computation time with varying batch size and epoch

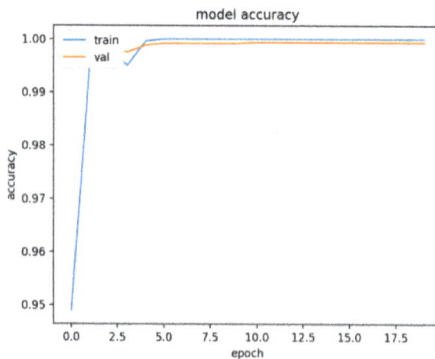

Figure 3: Plot of model accuracy with corresponding epoch

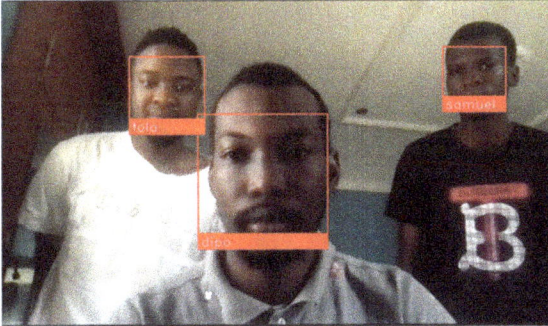

Figure 4: Detection of multiple registered users

Table 2: Face recognition test for five users

Users	Face Recognized				
1	True	True	True	True	True
2	True	True	True	True	True
3	True	True	True	True	True
4	True	True	True	True	True
5	True	True	True	True	True

Table 3: Time taken to train the model

	Epoch				
	0	5	10	15	20
Batch Size	**Training time (sec)**				
5	0	40.894	75.787	111.132	148.236
10	0	66.742	135.965	222.943	307.826
15	0	217.276	447.403	576.215	732.753
20	0	629.192	1088.779	1622.536	1892.426

The system was tested using five registered users, repeated five times for each. The system recognized all the users for each of the sessions. For the hand gesture, the system outputs the confidence level and a picture of the gesture captured (Figures 5(a-d)). The confidence level recorded varied between 98.15 and 99.92 throughout the experiment for the fist, palm, thumb with last finger, and thumbs up. These tests were carried out at the highest epoch value and batch number, which showed higher accuracy at the model training phase. However, when no epoch was set, the accuracy was below 10.2, whereas there was no accuracy below 98% when epoch was introduced. This depicts the importance of epoch in achieving accuracy in image recognition.

Figure 5a: identifies a palm

Figure 5b: identifies a fist

Figure 5c: identifies the thumb with last finger

Figure 5d: identifies the thumb up gesture

Conclusion

Facial recognition and hand gesture algorithm was developed and implemented for automatic gate control. Facial recognition and gesture control was achieved more accurately using high epoch and batch number. The model was trained with five gestures; fist, palm, thumb up, thumb, and last finger. The gesture recognition accuracy achieved through the epoch was maximum at 99.97 when both batch number and epoch were set to 20. However, when no epoch was set, the accuracy was below 10.2, whereas there was no accuracy below 98% when the epoch was introduced. This depicts the importance of epoch in achieving accuracy in image recognition. It was also discovered that the higher the epoch and the batch number the greater the accuracy, but the processing would require a very high processing unit.

References

[1] Turk M and Pentland A. Eigenfaces for recognition. Journal of Cognitive Neuroscience, 3(1):71-86, 1991. https://doi.org/10.1162/jocn.1991.3.1.71

[2] Bhatia R., Biometrics and Face Recognition Techniques. International Journal of Advanced Research, 3(5): 93-99, 2013.

[3] McConnel R. K., "Method of and apparatus for pattern recognition". United States of America Patent 4,567,610., January 1986.

[4] Deepan R., Rajavarman S. V. and Narasimhan K. Hand Gesture Based Control of Robotic Hand using Raspberry Pi Processor. Asian Journal of Scientific, 8(3): 392-402, 2015. https://doi.org/10.3923/ajsr.2015.392.402

[5] Chi-Man-Zhu P. and Wei H.-M. F., Real-Time Hand Gesture Recognition using Motion Tracking. International Journal of Computational Intelligence Systems, 4(10): 277-286, 2011. https://doi.org/10.1080/18756891.2011.9727783

[6] Sirovich L. and Kirby M. Low-dimensional procedure for the characterization of human faces. Optical Society of America, 4:519, 1987 https://doi.org/10.1364/JOSAA.4.000519

[7] Turk M and Pentland A. Eigenfaces for recognition. Journal of Cognitive Neuroscience, 3(1):71-86, 1991. https://doi.org/10.1162/jocn.1991.3.1.71

[8] Z. Sufyanu, F. S. Mohamad, A. A. Y. A. N. Musa and R. Abdulkadir. Feature extraction methods for face recognition. International Review of Applied Engineering Research (IRAER), 5(3): 5658-5668, 2016.

[9] Bhuyan M., Neog D. R. and Kar M. K. Fingertip Detection for Hand Pose. International Journal on Computer Science and Engineering (IJCSE), 4(3): 501-511, 2012.

[10] Elmezain M., Al-Hamadi A., Sadek S. and Michaelis B. Robust methods for hand gesture spotting and recognition using Hidden Markov Models and Conditional Random Fields. 10th IEEE International Symposium on Signal Processing and Information Technology,131-136, 2010. https://doi.org/10.1109/ISSPIT.2010.5711749

[11] Deepan R., Rajavarman S. V. and Narasimhan K. Hand Gesture Based Control of Robotic Hand using Raspberry Pi Processor. Asian Journal of Scientific, 8(3): 392-402, 2015. https://doi.org/10.3923/ajsr.2015.392.402

[12] P. Chi-Man-Zhu and H.-M. F. Wei. Real-Time Hand Gesture Recognition using Motion Tracking. International Journal of Computational Intelligence Systems, 4(10): 277-286, 2011. https://doi.org/10.1080/18756891.2011.9727783

[13] Tu Y.-J., Kao C.-C., Lin H.-Y. and Chang C.-C.. Face and Gesture Based Human Computer Interaction. International Journal of Signal Processing, Image Processing and Pattern Recognition, 8(9): 219-228, 2015. https://doi.org/10.14257/ijsip.2015.8.9.23

[14] Sharma R. P. and Verma G. K.. Human Computer Interaction using Hand Gesture. Eleventh International Multi-Conference on Information Processing-2015 (IMCIP-2015), India, 2015. https://doi.org/10.1016/j.procs.2015.06.085

[15] Januzaja Y., Lumaa A., Januzaja Y. and Ramaj V. Real Time Access Control Based on Face Recognition, International Conference on Network security & Computer Science, Antalya (Turkey), 2015.

AToMech1-2023 Supplement
Materials Research Proceedings 36 (2023) 8-15

Materials Research Forum LLC
https://doi.org/10.21741/9781644902790-2

Prediction of tool failure in metal hot extrusion process using artificial neural networks

Mosa Almutahhar[1], Ali Alhajeri[1], Rashid Ali Laghari[1,3],
Syed Sohail Akhtar[1,3], Usman Ali[1,2,4,a*]

[1]Department of Mechanical Engineering, King Fahd University of Petroleum & Minerals, Dhahran, 31261, Saudi Arabia

[2]Interdisciplinary Research Center for Advanced Materials, King Fahd University of Petroleum & Minerals, Dhahran, 31261, Saudi Arabia

[3]Interdisciplinary Research Center for Intelligent Manufacturing and Robotics, King Fahd University of Petroleum & Minerals, Dhahran, 31261, Saudi Arabia

[4]K.A. CARE Energy Research & Innovation Center at Dhahran, Saudi Arabia

[a]usman.ali@kfupm.edu.sa

Keywords: Hot Extrusion, Die Failure, H13 Steel, 6063 Aluminum, Artificial Neural Network

Abstract. The variation of tool performance and nonuniform process parameters in metal forming are some of the factors that complicate the tool life modeling and analysis of such processes. In this work, a brief discussion about machine learning in analyzing metal extrusion process as well as tool life modeling, and an implemented work of using machine learning to predict failure modes for H13 Steel die used in 6063 Aluminum hot extrusion process is presented. The analysis is conducted on a set of steel dies used in 6063 aluminum hot extrusion process. The data for the failed dies used in this work is collected from a local hot extrusion manufacturer. Using artificial neural network, the prediction of the die failure modes was modeled. Moreover, the model's accuracy and improvement recommendations are presented.

Introduction

Metal extrusion process is one of the most preferable forming processes since it enables the production of relatively complex cross sections with good surface finish [1] . In addition, it provides the flexibility of using the same press for different materials and final cross-sections by using different dies. Due to its excellent deformability as well as considerable strength properties, Aluminum is one of the most widely used working material in this process [2]. Extrusion can be classified as cold or hot process depending on the initial temperature of the billet to be extruded [1]. In hot extrusion, the working material is preheated above the recrystallization temperature [3], to allow processing with lower power requirements and extrusion time. Next, the preheated billet is forced within a chamber through a die to achieve the final end profile.

Along with the initial billet temperature and the extrusion force [4], there are multiple parameters related to the process, material, and geometry such as ram speed, friction at the interfaces, extrusion ratio, number of cavities, metallurgical condition of the billet and deformation characteristics of the tool and working material [5]. The relationship between the input parameters and the extrusion performance exhibits a non-linear correlation [6]. Therefore, it is important to identify an optimum parameter set. An optimum set of process parameters are required to guarantee a final product with acceptable quality while maintaining an efficient power consumption and a long die life [7].

Optimizing the extrusion process parameters and die life as well as studying the metal flow behavior and temperature distribution are the main areas of research in this process [8]. Among all

A I oMech1-2023 Supplement   Materials Research Forum LLC
Materials Research Proceedings 36 (2023) 8-15   https://doi.org/10.21741/9781644902790-2

the extrusion process research areas, die life optimization and modeling plays a vital role for the feasibility of the overall process since rapid die failure is a major factor that affects the profitability. For each die replacement, the costs of press down-time, costs of die elements fabrication or refurbishing, and costs of die handling and assembly are all increased [8]. Also, frequent die wear damages are reflected on the final product quality.

The prediction and optimization of die life could be done through a variety of systematic approaches, and they can be divided mainly into three groups as analytical-based models, experimental-derived models, and numerical computation models. Several experimental works have been done to investigate the die life in hot metal extrusion by these methodologies. Akhtar et. al. [9] used a statistical approach and proposed a regression model to predict the die life. A few researchers [10], [11] used Monte Carlo to predict the extrusion die fracture failure and to correlate the stochastic nature of various fatigue and wear related die variables to die life. Simulation based approaches are also used to predict the die life. Akhtar et al. [12] employed finite element (FE) simulations to analyze the extrusion process parameters. Li et al. [13] used FE simulations to study the wear failures in the extrusion dies. Redl et al. [14] also implemented a FE simulation to provide a numerical description of the mechanical and thermal loading during the extrusion process.

Current works from literature show a lack of appropriate tools in predicting the die failure modes in extrusion. Huge data sets required for accurate statistical approaches and time limits associated with FE simulations are some of the common issues. Machine learning approaches provide an alternative solution to predict die failure in hot extrusion. Artificial neural networks (ANN) is one such approach and has been shown to provide excellent predictive capabilities when compared to experimental and simulation results. ANN models give excellent predictive capabilities along with huge time savings when compared to numerical approaches [15]. Bhadeshia [16] has discussed applications of ANN models in materials science. Flow behavior at room and elevated temperatures has also been predicted using ANN models [17], [18]. Cyclic and static loading and damage has also been predicted using ANN models [19], [20]. In addition to these, machine learning approaches such as ANN, GA, and fuzzy learning have also been used in manufacturing processes [21]–[23]. For metal extrusion, several researchers have used ANN to predict the extrusion load [24]–[26]. While a few researchers have used machine learning approaches for die design [27], [28]. However, none of the studies focused on predicting failure modes for hot extrusion dies. In this work, a data driven ANN model is used to analyze die failures during hot extrusion process. The data was collected for H-13 steel dies used in extrusion of aluminum 6063 billets. Results show that ANN can be used for prediction of die failure modes.

Die Failure Mechanisms

Since the final extrudates profitability and quality are directly related to the die performance, comprehensive knowledge about the die failure mechanisms and modes is the key to achieve long die life. Tool-based and operation-based failure reasons are the two main categories for the most influencing factors on die life [8]. Die geometrical and material specifications as well as die manufacturing and surface treatment history are examples of the first category, where billet material and microstructure (recycled or pure) properties with equipment capability and operational settings are examples of the second category [8].

The commonly encountered failure mechanisms in metal extrusion due to those factors are fracture, wear, and deflection [29]. Fracture could be caused by fatigue loading or by thermal and mechanical loading [30]. Wear is induced by several reasons, such as adhesion, that gradual deterioration of the die surface [31]. Deflection occurs after heavy plastic deformation which may alter the shape of the die components [29]. During the service lifetime of the die, various damage types might overlap and act at the same time, which generates mixed modes of failure, with one significant factor. The correlated defects and failure types to the mentioned mechanisms are categorized and listed in Table 1.

Table 1: Failure types for each failure mechanism [29]

Failure Mode	Failure Types
Fracture (F)	Bearing chip-off (BCO), Corner crack (CC), Die broken/cracked (DB/DC), Bearing broken/cracked (BB/BC), Cavity broken (CvB), Tongue broken/cracked (TB,TC), Detail broken/Path broken/Tip broken/Screw broken/ (DtB)
Wear (W)	Bearing wash-out (BWO), Dimension change/oversize/overweight (DimC/OS/OW)
Deflection (D)	Cavity/die deflected (CvD/DD), Tongue deflected (TD)
Mixed mode (Mx)	Mixed mode (Mx)
Miscellaneous (Msc)	Bearing/cavity damage (BDm/CvDm), Corroded (Crd), Die Modified (DMod)
Mandrel failure(M)	Mandrel broken/cracked/deflected (MB/MC/MD), Web Cracked (WC)

Artificial Neural Network Model

Machine learning (ML) is one of the modern techniques that facilitates the understanding of the available data to draw practical conclusions. In theory, it is the field of study that intended to let computers learn without being explicitly programed where different methods and algorithms are implemented depending on the nature of the problem, number of variables and the suitable model [32]. The field of ML is highly diverse and categorizing its techniques can be achieved in different ways. One widely accepted approach is to separate the ML methods into three main domains as supervised learning, unsupervised learning, and reinforcement learning [33]–[36]. A good structuring of ML domains with some examples of their methods as well as their overlapping are presented in Fig. 1 [37].

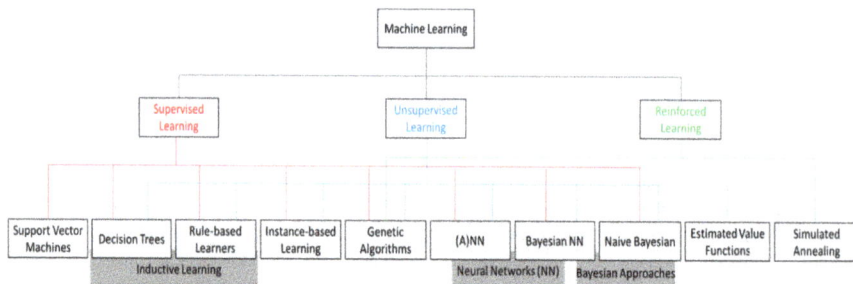

Fig. 1: Machine learning domains and examples [37]

In machine learning, the nature of problems usually depends on a knowledgeable source that states the assumptions and identification sets to train the algorithms [37] The most common techniques in supervised learning, as per the author's observation, are support vector machines (SVM), decision trees (DT), genetic algorithms (GA), and artificial neural network (ANN).

In this work, an ANN model was used to predict the failure modes for failed dies during hot extrusion. Data used in this work was gathered from Aluminium Products Company © (ALUPCO). The collected data for 135 failed H13 steel dies was sorted according to die type, profile, billet quality, nitriding history and failure history. A total of 89 data sets were used in this study. The failure modes were the outputs from the ANN model as shown in Table. 1 [29]. The final dataset used in this work was broken down into training (70%), validation (15%) and testing (15%) sets. It should be mentioned that the extrusion ratio (R) and Secondary Billet percentage (S%), were calculated as:

$$R = \frac{Area\ of\ original\ billet}{Area\ of\ extruded\ billet} \tag{1}$$

$$S\% = \frac{No.\ of\ secondary\ billets}{Total\ No.\ of\ extruded\ billets/die} \tag{2}$$

Results and discussion

The output of the artificial neural network (ANN) model was defined as the failure mode for the die. Several ANN models with different neurons and hidden layers were analyzed in multiple iterations to investigate the mean square errors (MSE) with the train, test and validation runs. Fig. 2 shows the iterations of the optimal ANN model. The graph shows the least MSE at 5th iteration. A breakdown of the error distributions for the optimal model is shown in Fig. 3. Error distributions show maximum predictions with near zero error showing reasonable accuracy of the proposed ANN model. A non-skewed distribution shows a well selected model but due to low accuracy, the model needs more work to get better predictions.

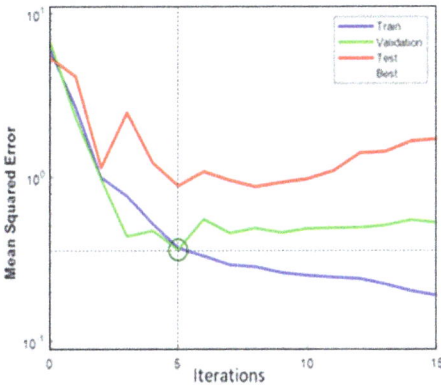

Fig. 2: Proposed ANN model accuracy as per MSE   Fig. 3: Proposed ANN model errors analysis

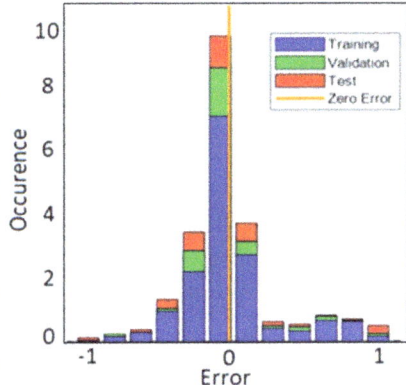

The model predictions along with experimental observations are shown in Fig 4. The linear line shows the correctly matched predictions. Results from the ANN model shows a training accuracy of 70% while testing and validation showed 50% accuracy respectively. However, the overall data set showed an accuracy of 67%. This low performance of the proposed model is highly associated with the choice of input parameters as well as the algorithm suitability for the current conditions. Involvement of more process parameters, such as billet temperature and ram speed, rather than the reliance on geometrical factors enhance the model training ability. Also, integrating other ML methods, such as SVM and GA, with the current ANN algorithm to add more flexibility and proper tuning could improve the model performance and give better results.

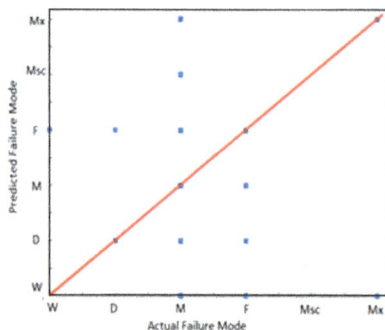

Fig. 4: Proposed ANN model validation

Conclusions

Metal extrusion process parameters are related to each other in a nonlinear manner. The optimization of these parameters is paramount to achieve extrudate with acceptable quality in an efficient power consumption and long die life. Predicting die failure mode is one such factor that that guarantees high productivity. In this work, a machine learning artificial neural network (ANN) model was used to predict the failure mode of extrusion dies. Compared to analytical, experimental, and numerical approaches machine learning based approaches provide a flexible and cost-efficient solution to this problem. A set of 89 data points were used to train, test and validate the ANN model. The input parameters to the model were the die type, cavities, and billet quality whereas failure mode was defined as the output. Results from the proposed model showed a 67% accuracy. It is proposed that the model predictions can be improved with additional parameters such as billet and die temperatures.

Acknowledgements

The authors would like to thank King Fahd University of Petroleum & Minerals for the financial support for this work. In addition, the authors would like to thank Interdisciplinary Center for Advanced Materials and Interdisciplinary Center for Intelligent Manufacturing and Robotics at King Fahd University of Petroleum & Minerals as well as K.A. CARE Energy Research & Innovation Center at Dhahran, Saudi Arabia. The authors would also like to thank the feedback and support from the members of Rapid Prototyping and Reverse Engineering Lab at King Fahd University of Petroleum & Minerals.

References

[1] R. Sathish, S. Vasanthakumar, S. Sasikumar, and M. Yuvapparasath, "A Metal Forming By Hot Extrusion Process," *International Research Journal of Engineering and Technology*, vol. 5, no. 7, pp. 1712–1714, 2018, [Online]. Available: www.irjet.net

[2] A. S. Chahare and K. H. Inamdar, "Optimization of Aluminium Extrusion Process using Taguchi Method," *IOSR Journal of Mechanical and Civil Engineering*, vol. 17, no. 01, pp. 61–65, Mar. 2017. https://doi.org/10.9790/1684-17010016165

[3] S. Chen and X. Jiang, "A Review of Modeling and Control for Aluminum Extrusion," in *International Conference on Artificial Intelligence and Computer Science (AICS 2016)*, 2016, pp. 565–573.

[4] M. M. Marín, A. M. Camacho, and J. A. Pérez, "Influence of the temperature on AA6061 aluminum alloy in a hot extrusion process," in *Manufacturing Engineering Society International*

Conference 2017, MESIC 2017, 2017, vol. 13, pp. 327–334.
https://doi.org/10.1016/j.promfg.2017.09.084

[5] S. N. A. Rahim, M. A. Lajis, and S. Ariffin, "Effect of extrusion speed and temperature on hot extrusion process of 6061 aluminum alloy chip," *ARPN Journal of Engineering and Applied Sciences*, vol. 11, no. 4, pp. 2272–2277, 2016, [Online]. Available: https://www.researchgate.net/publication/306215537

[6] S. Jajimoggala, R. Dhananjay, V. Lakshmi, and Shabana, "Multi-response optimization of hot extrusion process parameters using FEM and Grey relation based Taguchi method," in *International Conference on Advances in Materials and Manufacturing Engineering, ICAMME-2018*, 2019, vol. 18, pp. 389–401. [Online]. Available: www.sciencedirect.comwww.materialstoday.com/proceedings

[7] A. Medvedev, A. Bevacqua, A. Molotnikov, R. Axe, and R. Lapovok, "Innovative aluminium extrusion: Increased productivity through simulation," in *18th International Conference Metal Forming*, 2020, vol. 50, pp. 469–474. doi: 10.1016/j.promfg.2020.08.085.

[8] K. Lange, L. Cser, M. Geiger, and J. A. G. Kals, "Tool Life and Tool Quality in Bulk Metal Forming," *CIRP Ann Manuf Technol*, vol. 41, no. 2, pp. 667–675, 1992. https://doi.org/10.1016/S0007-8506(07)63253-3

[9] S. S. Akhtar, A. F. M. Arif, and A. K. Sheikh, "Influence of Billet Quality on Hot Extrusion Die Life and its Relationship with Process Parameters," in *Advanced Materials Research*, 2010, vol. 83–86, pp. 866–873. https://doi.org/10.4028/www.scientific.net/AMR.83-86.866

[10] S. Z. Qamar, "Fracture life prediction and sensitivity analysis for hollow extrusion dies," *Fatigue Fract Eng Mater Struct*, vol. 38, no. 4, pp. 434–444, Apr. 2015. https://doi.org/10.1111/ffe.12244

[11] S. Z. Qamar, A. K. Sheikh, A. F. M. Arif, M. Younas, and T. Pervez, "Monte Carlo simulation of extrusion die life," *J Mater Process Technol*, vol. 202, no. 1–3, pp. 96–106, Jun. 2008. https://doi.org/10.1016/j.jmatprotec.2007.08.062

[12] S. S. Akhtar and A. F. M. Arif, "Fatigue Failure of Extrusion Dies: Effect of Process Parameters and Design Features on Die Life," *Journal of Failure Analysis and Prevention*, vol. 10, no. 1. pp. 38–49, Feb. 2010. https://doi.org/10.1007/s11668-009-9304-4

[13] T. Li, G. Zhao, C. Zhang, Y. Guan, X. Sun, and H. Li, "Effect of Process Parameters on Die Wear Behavior of Aluminum Alloy Rod Extrusion," *Materials and Manufacturing Processes*, vol. 28, no. 3, pp. 312–318, Mar. 2013. https://doi.org/10.1080/10426914.2012.675536

[14] C. Redl *et al.*, "Investigation and Numerical Modelling of Extrusion Tool Life Time," in *7th International Tooling Conference*, 2006, pp. 589–596.

[15] U. Ali, W. Muhammad, A. Brahme, O. Skiba, and K. Inal, "Application of artificial neural networks in micromechanics for polycrystalline metals," *Int. J. Plast.*, 2019.

[16] H. K. D. H. Bhadeshia, "Neural Networks in Materials Science," *ISIJ Int.*, vol. 39, no. 10, pp. 966–979, 1999.

[17] Y. C. Lin, J. Zhang, and J. Zhong, "Application of neural networks to predict the elevated temperature flow behavior of a low alloy steel," *Comput. Mater. Sci.*, vol. 43, no. 4, pp. 752–758, 2008.

[18] Y. C. Lin, X. Fang, and Y. P. Wang, "Prediction of metadynamic softening in a multi-pass hot deformed low alloy steel using artificial neural network," *J. Mater. Sci.*, vol. 43, no. 16, pp. 5508–5515, 2008.

[19] W. Zhang, Z. Bao, S. Jiang, and J. He, "An artificial neural network-based algorithm for evaluation of fatigue crack propagation considering nonlinear damage accumulation," *Materials (Basel)*, 2016.

[20] A. Alshaiji, J. Albinmousa, M. Peron, B. AlMangour, and U. Ali, "Analyzing quasi-static fracture of notched magnesium ZK60 using notch fracture toughness and support vector machine," *Theoretical and Applied Fracture Mechanics*, vol. 21, 2022.

[21] H. Baseri, M. Bakhshi-Jooybari, and B. Rahmani, "Modeling of spring-back in V-die bending process by using fuzzy learning back-propagation algorithm," *Expert Syst Appl*, vol. 38, pp. 8894–8900, 2011.

[22] F. R. Biglari, N. P. O'Dowd, and R. T. Fenner, "Optimum design of forging dies using fuzzy logic in conjunction with the backward deformation method," *Int J Mach Tools Manuf*, vol. 38, no. 8, pp. 981–1000, 1998.

[23] F. R. Bittencout and L. E. Zarate, "Hybrid structure based on previous knowledge and GA to search the ideal neurons quantity for the hidden layer of MLP-Application in the cold rolling process," *Appl Soft Comput*, vol. 11, pp. 2460–2471, 2011.

[24] I. Zohourkari, S. Assarzadeh, and M. Zohoor, "Modeling and Analysis of Hot Extrusion Metal Forming Process Using Artificial Neural Network and ANOVA," in *10th Biennial Conference on Engineering Systems Design and Analysis*, 2010. [Online]. Available: http://proceedings.asmedigitalcollection.asme.org/pdfaccess.ashx?url=/data/conferences/esda201 0/72271/

[25] S. Nanne Saheb and S. Kumanan, "Modeling of Hot extrusion process using Artificial Neural Networks implanted with Genetic Algorithm," in *National Symposium on Advances in Metal Forming*, 2003.

[26] K. H. Raj, R. S. Sharma, S. Srivastava, and C. Patvardhan, "Optimization of Hot Extrusion using Single Objective Neuro Stochastic Search Technique," in *Proceedings of IEEE International Conference on Industrial Technology*, 2000. https://doi.org/10.1109/ICIT.2000.854248

[27] G. Zhao, H. Chen, C. Zhang, and Y. Guan, "Multiobjective optimization design of porthole extrusion die using Pareto-based genetic algorithm," *International Journal of Advanced Manufacturing Technology*, vol. 69, no. 5–8, pp. 1547–1556, Nov. 2013. https://doi.org/10.1007/s00170-013-5124-5

[28] S. Butdee and S. Tichkiewitch, "Case-Based Reasoning for Adaptive Aluminum Extrusion Die Design Together with Parameters by Neural Networks," in *Global Product Development - Proceedings of the 20th CIRP Design Conference*, 2011, pp. 491–496. https://doi.org/10.1007/978-3-642-15973-2_50

[29] A. F. M. Arif, A. K. Sheikh, and S. Z. Qamar, "A study of die failure mechanisms in aluminum extrusion," *J Mater Process Technol*, vol. 134, no. 3, pp. 318–328, Mar. 2003. https://doi.org/10.1016/S0924-0136(02)01116-0

[30] S. Z. Qamar, A. K. Sheikh, T. Pervez, and A. F. M. Arif, "Using Monte Carlo Simulation for Prediction of Tool Life," in *Applications of Monte Carlo Method in Science and Engineering*,

Prof. Shaul Mordechai, Ed. InTech, 2011, pp. 881–900. [Online]. Available: www.intechopen.com

[31] D. Lepadatu, R. Hambli, A. Kobi, and A. Barreau, "Tool Life Prediction in Metal Forming Processes Using Numerical Analysis," *IFAC Proceedings Volumes*, vol. 37, no. 15, pp. 287–291, 2004. https://doi.org/10.1016/s1474-6670(17)31038-8

[32] B. Mahesh, "Machine Learning Algorithms - A Review," *International Journal of Science and Research (IJSR)*, vol. 9, no. 1, 2020. https://doi.org/10.21275/ART20203995

[33] D. Y. Pimenov, A. Bustillo, S. Wojciechowski, V. S. Sharma, M. K. Gupta, and M. Kuntoğlu, "Artificial intelligence systems for tool condition monitoring in machining: analysis and critical review," *Journal of Intelligent Manufacturing*. Springer, 2022. https://doi.org/10.1007/s10845-022-01923-2

[34] C. C. Antonio, C. F. Castro, and L. C. Sousa, "Eliminating Forging Defects Using Genetic Algorithms," *Materials and Manufacturing Processes*, vol. 20, no. 3, pp. 509–522, 2005.

[35] J. Karandikar, "Machine learning classification for tool life modeling using production shop-floor tool wear data," in *47th SME North American Manufacturing Research Conference, NAMRC 47*, 2019, vol. 34, pp. 446–454. https://doi.org/10.1016/j.promfg.2019.06.192

[36] L. N. Pattanaik, "Applications of Soft computing tools in Metal forming: A state-of-art review," *Journal of Machining & Forming Technologies*, vol. 5, 2013.

[37] T. Wuest, D. Weimer, C. Irgens, and K. D. Thoben, "Machine learning in manufacturing: advantages, challenges, and applications," *Prod Manuf Res*, vol. 4, no. 1, pp. 23–45, Jun. 2016. https://doi.org/10.1080/21693277.2016.1192517

AToMech1-2023 Supplement
Materials Research Proceedings 36 (2023) 16-37

Materials Research Forum LLC
https://doi.org/10.21741/9781644902790-3

Ionanofluids: A review on its properties and thermal applications

Rahaf Almutairi[1,a*], Feroz Shaik[1,b] and Syam Sundar Lingala[1,d]

[1]Department of Mechanical Engineering, Prince Mohammad Bin Fahd University, Al Khobar, Kingdom of Saudi Arabia

almutairirahafn@gmail.com[a], ferozs2005@gmail.com[b], sslingala@gmail.com[c]

Keywords: Ionanofluids (INFs), Nanoparticles (NPs), Ionic Liquids (ILs), Heat Transfer Fluids (HTFs), Thermophysical, Thermal Applications

Abstract. The dispersion of nanoparticles (NPs) into ionic liquids results in ionanofluids (INFs) (ILs). INFs are regarded as the newest type of heat transfer fluids (HTFs). INFs are discovered to improve the fluids' thermophysical characteristics at high temperatures with negligible vapor pressure. The preparation techniques and theoretical models used to calculate the physical characteristics of ionofluids, including their density, heat capacity, thermal conductivity, and thermal applications, are summarized in this work.

Introduction

Nanotechnology focuses on creating and using objects with organizational characteristics between individual molecules and around 100 nm, where special properties emerge in contrast to bulk materials. To build tailored nanostructures and gadgets for particular uses, it entails changing molecules and atoms. Nanotechnology is growing and enabling technology for the twenty-first century, alongside the already well-established sectors of information technology and biotechnology. The convergence of knowledge in physics, chemistry, biology, materials science, and engineering at the nanoscale is the cause, and matter control at the nanoscale is significant for almost all technologies.

Because manufacturing of nanoparticles is a key component of nanotechnology, specific features are achieved at the nanoparticle, nanocrystal, or nanolayer level. The assembly of precursor particles and related structures is the most popular technique for producing nanostructures. People have been producing nanoparticles empirically for thousands of years, for as by producing carbon black [1]. Ionanofluid (INF) is defined as a nanodispersion of metallic, polymeric, inorganic, or organic solid nanoparticles dispersed in an ionic liquid (IL) [2,3]. Since Nieto de Castro's proposal of such a system, INFs have received considerable scholarly interest. Ionic liquids (ILs) are described as liquids with organic cations and organic/inorganic anions that are at room temperature. They can be used as heat transfer fluids (HTFs) because of their good thermal characteristics [2,4]. The characteristics of vapor pressure and volatility are deemed unimportant for ILs. In addition to being used as HTFs, ILs have a variety of additional uses. For instance, they are frequently employed as solvents in chemical, gas processing, coal handling, and pharmaceutical factories, among other facilities [5–10].

Nanofluids have a lot of potential for applications involving heat transmission. They are made up of tiny nanoparticles in volume and weight fractions together with base fluids like water and glycol. When operating temperatures range from -40 to 400 °C, ILs have more advantages when used as a continuous phase in nanofluids than the usual base fluids. INFs have great dissolving power for a variety of organic and inorganic substances, considerable chemical and stable thermal characteristics, and little vapor pressure [11–15]. The scientific discipline of rheology gains substantially from nanotechnology. It offers several useful tools to look at how various physical and chemical circumstances affect ionanofluid properties. To validate materials prior to manufacturing and usage on a wider range, design the appropriate machineries, and improve the

 AToMech1-2023 Supplement   Materials Research Forum LLC
Materials Research Proceedings 36 (2023) 16-37   https://doi.org/10.21741/9781644902790-3

performance, such complex system necessitate in-depth rheological analysis [16,17]. Viscosity is the most important parameter to consider when looking at rheological parameters. Convective heat transfer, pressure loss, and required pumping power are all dependent on viscosity [18–20]. It is due to the requirement for optimizing internal resistance and heat transport capacities. Internal resistance increases as a fluid's viscosity increases [21].

Dynamic (also known as oscillatory) shear testing and steady shear experiments are two methods used to research rheological behavior. The majority of literature data follows the steady-shear material function, and steady shear flow is simple to manufacture. The material is subjected to a sinusoidally varying deformation in dynamic shear flow [22–24]. By employing steady shear measurements depends on viscosity or shear stress are studied. While the modulus rigidity (G") can be determined using dynamic shear tests as a function of temperature, time, or oscillation frequency. For example, moduli G' and G" show the material's "elastic" and "viscous" characteristics "properties, or stored and lost energy, during sample deformation, respectively [25,26]. Both rheological methods have benefits and drawbacks. The steady shear tests have several limitations in terms of sample component migration and slippage even if they require less complicated equipment. The non-destructive nature of oscillatory shear tests is due to the tiny strain/stress amplitude. They are, nevertheless, inappropriate for real-world processing applications [27,28].

The literature claims that nanodispersions exhibit complicated rheological characteristics [29]. There has already been a large amount of structured and reviewed research on the rheological properties of conventional nanofluids [6,18,20,30-44]. However, Shakeel et al. [45].'s previous research on viscosity of ionanofluids only included one review publication. Unfortunately, given of the review's broad topic scope, which encompassed the ionanofluids gels. The review by Vignarooban et al. [46] explored the thermal properties and stability of different ILs-based HTFs, including molten salts, steam, organic fluids, thermal oils, liquid metals, and air. For usage in parabolic trough collectors, Malviya et al. [47] investigated a number of ILs-based HTFs, including sodium salt, solar salt, solar grade oil, etc. Other ILS-based HTFs, including as Therminol VP-1, Syltherm 800, Solar salt, Hitec XL, and liquid sodium, have also been the focus of a theoretical investigation for Malaysian circumstances by Zaharil and Hasanuzzaman [48]. Conroy et al. [49] developed a theoretical model to investigate the hydraulic flow properties of liquid sodium, molten salt, and lead-bismuth in a concentrated solar power (CSP) receiver.

Trabelsi et al. [50] performed an optimization analysis of a solar parabolic trough power plant using the simulation software SAM, with an emphasis on the field size, storage system, and HTFs. They came to the conclusion that Therminol VPI and synthetic oil are less economically sound than molten salt. The thermal and physical properties of liquid sodium and Hitec (a blend of KN, NaN, and NaN) were assessed by Boerema et al. It has been established that future CSP can use liquid sodium in place of molten salts. Although there has been a lot of research in this field over the last ten years, more work is still required. This paper's major goal is to present a thorough analysis of ionanofluid in terms of its preparation processes, varieties, and properties. Additionally, to examine the effects of various parameters, such as thermal conductivity, viscosity, density, and specific heat, on the thermal transport characteristics of ionanofluid. The thermal uses of ionanofluids and potential thermal applications are covered in this review paper.

Preparation methods for Ionanofluids
In their review, Azmi et al. [40] discussed the processes for creating and purifying ionic liquids. One-butyl-3-methylimidazolium bis(trifluoromethyl) sulfonylimide, one-butyl-3-methylimidazolium ethylsulfate, one-butyl-3-methylimidazolium tetrafluoroborate, and one-hexyl-3-methylimidazolium tetrafluoroborate are the most often utilized ionic liquids as base liquids. Ionic liquids excel in several areas, including thermal stability, a wide operating temperature range, low freezing point, low flammability, low vapor pressure, high thermal

AToMech1-2023 Supplement   Materials Research Forum LLC
Materials Research Proceedings 36 (2023) 16-37   https://doi.org/10.21741/9781644902790-3

conductivity, and high heat capacity. The biggest drawback of employing ILs is their high viscosity, which increases the expense of pumping. By dispersing different weight/volume fractions of nanomaterials in the matching base ionic liquids, ionanofluids are created. Ionanofluids, for instance, were created by dispersing different weight percentages of Multiwall Carbon Nanotubes (MWCNT) in a variety of ionic liquids, then sonicating them for enhanced nanotube dispersion [52, 53].

Types of Ionanofluids
The creation of mixes of dispersed nanoparticles in predetermined base liquids was the basis for Choi and Eastman's initial mention of nanofluids [54]. Nanofluids are categorized into four primary classes based on the different types of nanoparticles that were employed in their synthesis: metal-based, metal oxide-based, carbon-based, and hybrid/mixed metal-based. To create nanofluids, these nanoparticles are suspended in base fluids such as water, methanol, ethylene glycol, kerosene, and transformer oil. The stability and physical properties of nanofluids must be taken into account when choosing them for any application. Nieto de Castro et al. [2] produced what are known as "ionanofluids" for the first time by dispersing nanoparticles within ILs. On the basis of their successes in improving thermal conductivity, other research projects were published in the synthesis of ionanofluids using the same techniques as nanofluids.

Properties of Ionanofluids
Because of their high thermal stability and low vapor pressure, ionic liquids offer a great deal of potential as heat transfer fluids. Additionally, they have an excellent thermal conductivity and large volumetric heat capacity (like more conventional HTFs like Dowtherm MXTM, Syltherm 800TM, and engine oil). Table 1 displays reference values for the thermophysical characteristics of various commonly used heat transfer fluids and ionic liquids at a temperature of 40 °C [55, 56].

Table 1. Reference values of various widely used heat transfer fluids and ionic liquids' thermophysical properties at a temperature of 40 °C

Liquids	l [W/m·K]	h [mPa·s]	Cp [J/kg·K]	r [kg/m^3]
Water	0.631	0.653	4,179	992
Ethylene glycol	0.256	10.37	2,520	1,100
Engine oil	0.148	568.00	2,000	880
1-butyl-3-methylimidazolium bis(trifluoromethyl) sulfonylimide	0.116	28.50	1,372	1,423
1-butyl-3-methylimidazolium ethylsulfate	0.178	50.00	1,615	1,226

The higher thermal conductivity of the nanoparticles in contrast to any base fluid has been attributed to the heat-transfer characteristics of nanofluid. Since thermal conductivity is the most important feature in this context, it is well described [57–61] how to improve the thermal conductivity of nanofluids. Other essential qualities, such as density, heat capacity, and viscosity, can be changed by the addition of nanoparticles. The effects of the nanoparticles are now more complicated, and it is still need to thoroughly examine all of the mechanisms. It is commonly known that when a fluid's temperature increases, its properties can change significantly. If we exclude molten salt-based nanofluids, there are relatively few research and observations of the thermophysical properties of nanofluids at temperatures beyond 100°C [62]. In this section, the characteristics of ionanofluids and the findings of literature searches in high-temperature regions are reviewed.

Thermal conductivity of Ionanofluids
The fact that nanofluids transmit heat more efficiently than base fluids is well-known. It has served as the subject of several literary analyses. Thermal conductivity was the subject of a thorough and insightful review by Sobti and Wanchoo [63]. Ionic liquids (ILs) have gained significant attention as environmentally friendly solvents, heat carriers, and electrolytes. Cations and anions can be made to have specific thermophysical properties and functions by changing their species. Understanding the basic thermophysical characteristics of ILs, such as their densities, viscosities, and thermal conductivities, is necessary to create ILs with the best possible thermophysical qualities. Tomida [64] presented the experimental results for the thermal conductivity of the pure IL components as well as methods for calculating the thermal conductivity of IL using correlations.

Additionally, there has recently been a lot of interest in the thermal conductivities of ionanofluids, which are made up of nanoparticles dispersed in an IL. In a study of the thermal conductivities of nanofluids with carbon nanotubes dispersed in ILs, Castro et al. provided the first description of the thermal conductivities of ionanofluids in 2010. The thermal conductivities of ILs have been shown to be rather low, matching those of ethanol and methanol. As a result, it was hypothesized that distributing nanoparticles within an IL would increase its thermal conductivity [65–67]. The impact of volume and temperature on the thermal conductivity of ionanofluids is covered in the section that follows. It also discusses the theoretical frameworks that were used to forecast the thermal conductivity of ionanofluids.

Effect of volume on thermal conductivity of Ionanofluids
Nanofluids were projected to offer superior properties over microfluids due to their large surface area to particle volume ratios. The Hamilton-Crosser model (Eq. 1) and the thermal conductivity of nanofluids are widely known to accord [68].

$$\frac{k_{eff}}{k_0} = \frac{k_p+(n-1)k_0+(n-1)(k_p-k_0)\phi_p}{k_p+(n-1)k_0-(k_p-k_0)\phi_p} \tag{1}$$

Where n is the particle shape parameter, k_p is the volume fraction of the dispersoid, and k_0, k_{eff}, and k_p are the thermal conductivities of the dispersion medium, nanofluid, and dispersoid, respectively, in [W/ m.K]. Fig. 1 depicts the projected rise in k value through Hamilton-Crosser model for Ag, Al_2O_3, $BaTiO_3$, nanoparticles are used as the dispersoid in 1-butyl-3-methylimidazolium tetrafluoroborate ([BMIM] [BF_4]) as a typical IL.

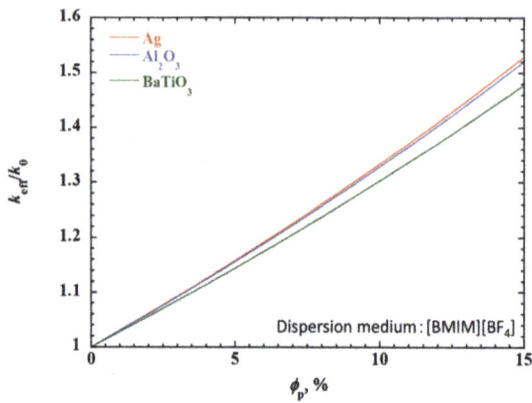

Figure 1. Relationship between the volume fraction of nanoparticles and the improvement in thermal conductivity based on the Hamilton-Crosser model at 298 K [64].

Table 2 shows k of each dispersed that was evaluated. The results show that the increase in k that occurs when the volume percentage is around 15% (n = 3) does not very much, even when dispersions have thermal conductivities that are significantly variably. This explains by diluting spherical nanoparticles in an IL, the thermal conductivity of the dispersed has little impact on the k resulting ionanofluid. In order to further increase the k of an ionanofluid containing spherical nanoparticles, the volume fraction of the dispersion must be increased; however, this has an impact on flowability. When using an ionanofluid as a heating medium, the minor volume proportion of nanoparticles must be used to enhance k. To dramatically increase the rate of k augment with a low nanoparticle concentration, it is essential to increase the particle diameter, n. Increasing n while lowering the slope is attainable by selecting a material with a high aspect ratio.

Table 2. Thermal conductivities of dispersoids [64].

Dispersoid	k_p, [W(mK)]
Ag	428
Al_2O_3	36
$BaTiO_3$	6

Effect of temperature on thermal conductivity of Ionanofluids
Studies have shown that temperature has a big impact on k data. The greater the impact, the higher the temperature. An increase in Brownian motion is the primary explanation for this effect [69–70]. But according to various research [71–73], the thermal conductivity of nanofluids closely mirrored the base fluid's temperature dependency. Fig. 2 displays temperature-dependent thermal conductivity data for ionanofluids containing the same concentration of MWCNT and SWCNT (1 wt%) in both water and [C_4mim] [NTf_2] and [C_2mim] [$EtSO_4$] [74]. At normal temperature, other [C_2mim] [$EtSO_4$] based ionanofluids showed an increase in thermal conductivity of just 8.5%,

Materials Research Forum LLC
https://doi.org/10.21741/9781644902790-3

whereas $[C_4mim]$ $[NTf_2]$ based ionanofluid showed an increase in k of up to 35.5% over its base ionic liquid.

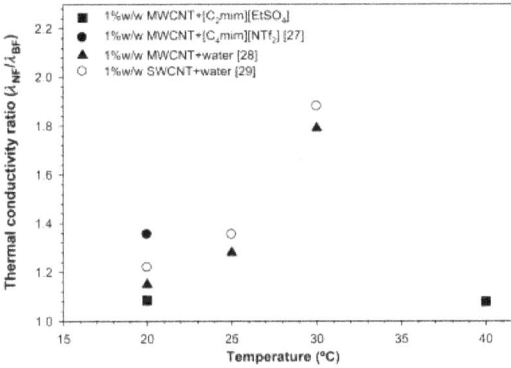

Figure 2. Effect of temperature on k of CNT loaded nanofluids and Ionanofluids [75]

At room temperature, $[C_2mim]$ $[EtSO_4]$ was found to have k of 0.128 W/(m. K). However, the MWCNT/water nanofluid k data published by Ding et al. [75] and the SWCNT/water nanofluid data presented by Amrollahi et al. [76] both exhibited increases in thermal conductivity of about 15% and 22%, respectively, at the same temperature and concentration. Fig. 2 further demonstrates that k of $[C_2mim]$ $[EtSO_4]$ based nanofluid is temperature independent, but the thermal conductivity of CNT-nanofluids increases noticeably as temperature rises.

Models to predict the thermal conductivity of Ionanofluids
Since the publication of Maxwell's treatise, various classical models have been constructed to estimate the effective thermal conductivity of suspensions of solid particles. Most researchers discovered that these classical models, including those credited to Maxwell [77] and Hamilton-Crosser [78], were not able to anticipate the anomalous thermal conductivity of nanofluids because they were designed for micro- or milimeter-sized dispersed particles. As a result, several theoretical investigations have been conducted over the past ten years to comprehend the heat transfer mechanism and create models that can precisely predict the effective thermal conductivity of nanofluids [79]. The majority of theoretical investigations, however, were restricted to spherical nanoparticles and were not well received.

Murshed et al. [80] developed a model for the prediction of k of cylindrical nanoparticle based nanofluids by considering the effects of particle size and the interfacial layer at the particle/fluid interface (nanolayer). In this study, both the traditional HC model and the more recent model developed by Murshed et al. are used to predict the effective k of ionanofluids containing MWCNT. The HC is modified version of Maxwell's model to include a form component that takes spherical and non-spherical particles' effective k into consideration. The volume ratio, the shape of the dispersed particles, and the k of the solid and liquid phases all play a role in their model. The form of the HC model is represented by Eq. 2.

$$\lambda_{eff} = \lambda_f \left[\frac{\lambda_p + (n-1)\lambda_f - (n-1)\phi_p(\lambda_f - \lambda_p)}{\lambda_p + (n-1)\lambda_f + \phi_p(\lambda_f - \lambda_p)} \right] \tag{2}$$

Where ϕ_p is the particle volume fraction, n is the form factor n = 3 for spherical particles and n =6 for cylindrical particles, and λ_f and λ_p are the thermal conductivities of the base fluid and nanoparticle, respectively. This idea states that spherical particles show a smaller increase in thermal conductivity than non-spherical ones. It is noted that for spherical particles, the Maxwell model can replace the Hamilton-Crosser model. On the other hand, Eq. 3 represents the model created by Murshed et al. for the effective k suspensions of cylindrical nanoparticles (nanofluids), which accounts for the effects of particle size, concentration, and interfacial nanolayer.

$$\lambda_{\text{eff-nf}} = \lambda_f \frac{\phi_p \omega (\lambda_p - \omega \lambda_f)[\gamma_1^2 - \gamma^2 + 1] + (\lambda_p + \omega \lambda_f)\gamma_1^2 [\phi_p \gamma^2 (\omega - 1) + 1]}{\gamma_1^2 (\lambda_p + \omega \lambda_f) - (\lambda_p - \omega \lambda_f)\phi_p [\gamma_1^2 - \gamma^2 - 1]} \tag{3}$$

Where $\omega = \lambda_{lr}/\lambda_f$, $\gamma = 1 + (t/r_p)$, $\gamma_1 = 1 + (t/2r_p)$, r_p is the particle's radius, t is the interfacial layer's thickness, and λ_{lr} is the interfacial layer's thermal conductivity.

According to experimental and analytical investigations, the thickness of a nanolayer produced at the interface between nanoparticles and fluids is thought to be 1 nm; however, neither theoretical nor practical methods can be used to evaluate the thermal conductivity of such a nanolayer. However, the sequence and orientation of the fluid molecules that are absorbed on a nanoparticle surface led to an intermediate value of thermal conductivity for the nanolayer, i.e., $\lambda_{fr} < \lambda_{lr} < \lambda_p$. The interfacial layer thermal conductivity is therefore expressed as by $\omega > 1$, where > 1 is an empirical value that depends on the arrangement of fluid molecules at the interface as well as the composition and surface chemistry of nanoparticles.

Viscosity of Ionanofluids
The addition of particles to a liquid alters the viscosity of the combination. This effect might not be noticeable for very small particle volume fractions. As the volume fraction rises, though, the mixture's viscosity could noticeably increase. Since viscosity is a crucial characteristic that impacts friction and pumping power/pressure drop, it has been the subject of several investigations. The rheological behavior of nanofluids has been the subject of some discussion [41]. There appears to be no correlation between rheological behavior and particle concentration, and the behavior appears to vary depending on the material, according to a recent review by Okonkwo et al. [59].

Wittmar's work [81] measured the apparent viscosity of INFs having three various kinds of externally purchased and internally produced TiO_2 nanoparticles distributed in three various ILs with various alkyl chain lengths: [Emim][BF_4], [Bmim][BF_4], and [Hmim][BF_4]. The amounts were 0.05, 0.1, 0.5, and 1 weight percent. The experiments were performed with shear rates ranging from 0.1 to 1000 s^{-1} at a temperature of 25 °C. The outcomes demonstrated that the length of the alkyl chain on the IL cation improved the viscosity and non-Newtonian characteristics of the INFs (Fig. 3). Longer alkyl chains made INFs more viscous because bigger cations had less rotational flexibility and greater van der Waals interactions [82–85]. Additionally, they inhibited Brownian motion from causing the nanoparticles to aggregate, which normally raised the stability of nanodispersion.

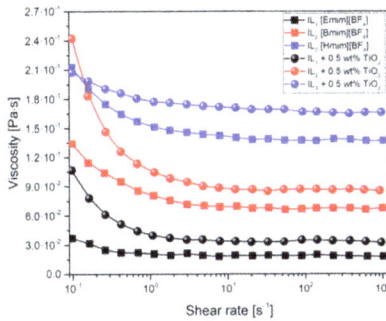

Figure 3. The figure depicts apparent viscosity as a function of shear rate for pure ILs and INFs that contain 0.5 weight percent of TiO$_2$ nanoparticles and have varying alkyl chain lengths on the IL cation. Data from [82] was used to make the figure.

Wittmer and Uibricht [86] looked viscosity of TiO$_2$//[Emim][NTF$_2$] and [Emim][BF$_4$] ionanofluids and found higher viscosity. In contrast to prolonged treatments, shorter ultrasonic treatments for hydrophilic IL resulted in Newtonian behavior of the INFs. But in both cases, a longer ultrasound session (>1 h) had little to no effect on the stability of INFs. The stability and rheology of nanodispersion appeared to be significantly influenced by the hydrophilicity of IL. It occurs because the hydroxyl groups of IL on the surface of nanoparticles and within the IL itself form hydrogen bonds. [Emim][BF$_4$] produced more stable and viscous nanodispersions than [Emim][NTF$_2$]. It might be connected to the [87] less adaptable chains [BF$_4$] ionanofluid. These observations are in line with other experimental results reported in the literature [88,89], which show that base ILs with the same cation but various anions display the sequence shown below as their viscosity rises: [PF$_6$]$^-$ > [BF$_4$]$^-$ > [NTf$_2$]$^-$ > [DCA]$^-$ > [TCB]$^-$ > [TCM]$^-$.

Gao et al. [88] observed the enhanced viscosity of [Bmim][BF$_4$] ionanofluids. Ueno et al. [89] examined viscosity of INFs dispersed in [Emim][NTF$_2$] and [Bmim][BF$_4$] that had hydrophilic and hydrophobic SNP loadings of 5, 8, 10, and 15 weight percent. The shear rates used for the rheological measurements ranged from 0.1 to 1000 s^{-1} at a temperature of 25 °C. At the lowest shear speeds, the INFs with hydrophilic SNPs and [Emim] [NTF$_2$] displayed significant shear thinning behavior and had a high apparent viscosity. The shear flow's destruction of the intermolecular physical connections in the flocculated silica networks is most likely what caused the non-Newtonian characteristics. The lack of silica network structure was suggested by the fact that hydrophilic SNP nanodispersions in [Bmim][BF$_4$] lower responsive to shear rate (Fig. 4). However, a distinct non-Newtonian behavior was showed that nanoparticles may have flocculated.

Materials Research Forum LLC

https://doi.org/10.21741/9781644902790-3

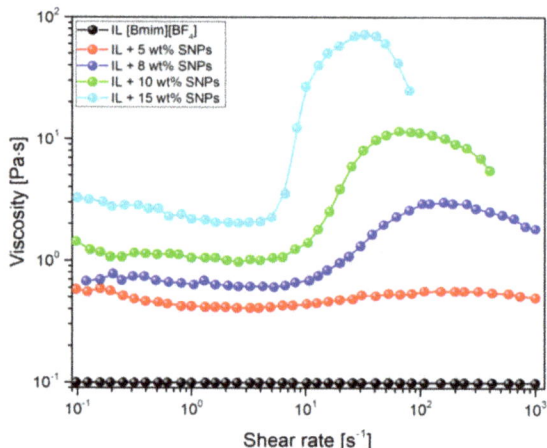

Figure 4. Viscosity of SNP ionanfluids (25 °C) [89,90].

Novak and Britton [91] investigated the shear stress and apparent viscosity of INFs with hydrophilic SNP loadings of 5 and 15 wt%, as well as two ILs, hydrophobic $[P_{6,6,6,14}]$ $[NTF_2]$ and hydrophilic [Bmim] $[BF_4]$. The experiments were performed with shear rates ranging from 0.05 to 1000 s^{-1} at a temperature of 25 °C. According to the published experimental data, Newtonian fluids were all that pure ILs could be at low shear rates (<10 s^{-1}), where the addition of 15 weight percent of SNPs to [Bmim] $[BF_4]$ produced Newtonian behavior, whereas at higher shear rates (>10 s^{-1}) exhibits non-Newtonian pseudoplastic characteristics. These results were somewhat in line with those of Ueno et al. [89]. The INF also demonstrated a more complex shear-banding behavior with five wt% of SNPs in $[P_{6,6,6,14}][NTF_2]$.

Zhang et al. [92] evaluated the stress and viscosity of Graphene/[Emim][Ac] ionanofluids and observed an enhanced viscosity. The self-lubricating characteristics of GNPs are most likely what caused this scenario [93]. The viscosity of GNPs/[Hmim] [BF 4] IL was measured by Liu et al. [94] and observed higher viscosity 170 °C, when it eventually plateaued at 6.3 mPas. Additionally, it was found that at the same temperature viscosity of the INFs was marginally lower. As was previously indicated, this phenomenon was assumed to be caused by the self-lubrication of GNPs.

The viscosity of [Hmim] $[BF_4]$/SiC ionanofluids was analyzed by Chen et al. [95]. Aggregates of nanoparticles most likely caused the increase. In the nanodispersions, free ions can be captured by aggregation of nanoparticles. The viscosity of MWCNTs/pure [Emim][DEP] IL were examined by Xie et al. [96]. The viscosity rose linearly when MWCNT concentration was raised. Additionally, it was found that when the base liquid's water content increased, viscosity drastically dropped (by as much as 75%). (Fig. 5).

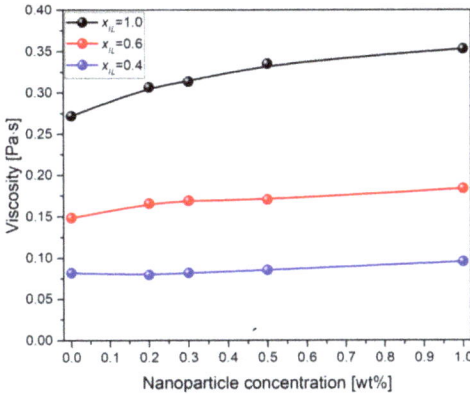

Figure 5. Viscosity of MWCNTs/[Emim][DEP] ionanofluids [96].

Density of Ionanofluids

When applied to a continuum of non-interacting particles, the traditional definitions of density and specific heat capacity give rise to straightforward equations for the mixture equivalents [58,97]. In other words, the mixture density, ρ_{nf} (kg/m^3). represents an average of the volume-based component densities.

$$\rho_{nf} = \alpha_p \rho_p + \alpha_{bf} \rho_{bf} \tag{4}$$

$$c_{nf} = \phi_p c_p + \phi_{bf} c_{bf} \tag{5}$$

The terms "nanofluid property," "particle property," and "base fluid property," respectively, are denoted by the subscripts nf, p, and bf. The density and volume fraction of the specified component are, respectively, and. The specific heat capacity of the mixture, and c_{nf} (J / (K kg)), is an average of the specific heat capacities of the various components depending on mass fraction [98]. Where $\phi_p = \alpha_i \rho_i / \rho_{nf}$ is the component mass-fraction. More complex models may be used to determine the component densities and specific heat capacities. However, these models might lose precision for small particle sizes. For instance, experimental findings by Sharifpur et al. [99] demonstrate that Eq. (4) miscalculates the density. They assume that the generally used density model ignores the effect of the gap between the nanoparticles and the base fluid because of the nano layer on the particle surface. They suggest a new theory to explain this nano layer:

$$\rho_{nf,new} = \frac{\rho_{nf}}{\alpha_{bf} + \alpha_p (1 + \frac{d_{nl}}{r_p})^3} \tag{6}$$

Where d_{nl} stands for the thickness of the nanolayer and r_p stands for the average particle diameter. Additionally, they demonstrate that their model has a superior fit with the detected experimental results. Hentschke [100] asserts that the improvement might be predominantly attributable to a different factor than the commonly believed nanolayer effect. He proposes a different theory instead, one that depends on longer-range interactions between the nanoparticles in the surrounding liquid. Adding 0.1 weight percent of SiO$_2$ and Al$_2$O$_3$ nanoparticles to Therminol 66 at temperatures between 280 and 320 °C did not appreciably change the fluid's density or viscosity compared to the base substance, according to a recent study by Safaei et al. [101]. There

AToMech1-2023 Supplement   Materials Research Forum LLC
Materials Research Proceedings 36 (2023) 16-37   https://doi.org/10.21741/9781644902790-3

is currently no comprehensive literature on the density and high temperature heat capacity of nanofluids.

Even if there is evidence of nanoparticle effects, such as the nanolayer, the fundamental connections in E. qs. (4) and (5) may still hold true at high temperatures, as the densities of the base fluid and the nanoparticles or nanoparticle material do typically rely on temperature [102].

$$\rho(T) = \sum_{i=0}^{2} a_i T^i \tag{7}$$

Where a_i : the regression parameter (using the least-square method), δa_i is the standard uncertainty, while the fit was described using the coefficient of determination, R^2. An experimental thermophysical characterization was carried out by Oster et al. [103]. The density was measured as a function of temperature, T = (298.15-363.15) in accordance with Eq. 7 using an Anton Paar DMA 4500M densitometer (calibrated on ultrapure degassed water and dry air, relative standard uncertainty of density $u_r(T) = 0.1\%$, relative standard uncertainty of temperature $u_r(T) = 0.01$ K, approximately 1.5 cm^3 of sample volume, three independent measurement repetitions.

The densities of multi-walled carbon nanotubes, boron nitride, and graphite nanoparticles were measured before determining the Ionanofluids density [104]. Despite not having been previously reported, those for mesoporous carbon were computed using the empirical formulas as stated in Eq. 8, in a manner akin to that followed for carbon nanotubes, boron nitride, and graphite [105]:

$$\rho_{NP} = \frac{\rho_{IL+NP} - \rho_{IL}(1 - w_{NP})}{w_{NP}} \tag{8}$$

The subscripts IL and NP stand for ionic liquids and nanoparticles, respectively, where w stands for the mass fraction. Eq. 8 can be used to evaluate density.

Heat capacity of Ionanofluids

Although the specific heat capacity may also decrease, adding nanoparticles usually improves a fluid's thermal conductivity. Its cooling capacity is severely limited. The alteration or improvement of thermophysical properties must be balanced in order to use a nanofluid during the application. For effective and clean heat transfer systems, higher heat-capacity heat transfer fluids are required. Researchers have suggested a variety of techniques, such as increasing surface energy, for improving specific heat capacity. They raised the heat resistance at the interfaces between the liquid molecules surrounding the nanoparticles and themselves, resulting in the formation of a semisolid liquid layer. In a recent study using molecular dynamics simulations, Carrillo-Berdugo et al. [104] attribute some of the specific-heat enhancement of nanofluids to strong chemisorption interaction of the fluid molecules at the nanoparticle surfaces.

Understanding specific heat capacity is essential for figuring out other heat transfer factors, flow characteristics, and enthalpy estimations when modelling different processes, it is well acknowledged. Fig. 6 demonstrates that MWCNT-ionanofluids have a higher specific heat capacity than base ionic liquids (i.e., $[C_4mim]$ $[PF_6]$) from room temperature to 115°C, and that this specific heat capacity grows with temperature between 60°C and 90°C [106]. The most noteworthy feature of these findings is that there is a dome-shaped spike in the specific heat capacity within a specific temperature range (60-110° C), independent of MWCNT loading (peak increase of 8% compared with base ionic liquid). It is yet unknown why these puzzling consequences exist.

Materials Research Forum LLC

https://doi.org/10.21741/9781644902790-3

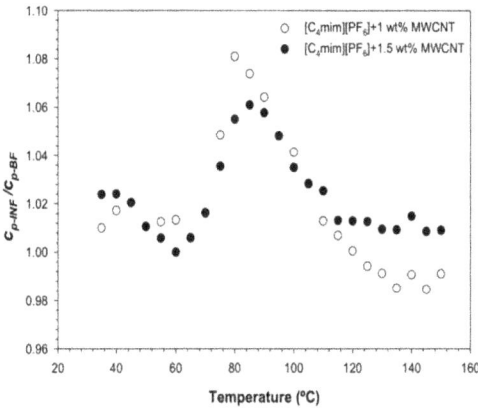

Figure 6. The specific heat of Ionanofluids [106].

Oster [103] measured the isobaric heat capacity, c_p by using differential scanning calorimeter (DSC) apparatus, Q100 TA Instruments, in the temperature range of 298.15-363.15 K (modulated differential scanning calorimetry technique, MDSC, calibrated on synthetic sapphire, CAS: 1317-824, TA Instruments, ultrapure in accordance with the reference standards NIST SRM 720) and checked with 1hexyl31hexyl3methylimidazolium bis(trifluoromethylsulfonyl)imide as a classified ionic liquid NIST standard [107]. It was performed under nitrogen gas flow = 50 cm^3 min^{-1}, heating rate dT/dt = 3 K min^{-1}, amplitude = ±0.5 K, modulation period = 60 s . The calibration and post-calibration method findings determined that the standard temperature uncertainty was 0.01 K, and the relative standard uncertainty of measurement was $u_r(T)$ = 3%. With a 5% repeatability, three independent measurement repeats were carried out. DSC equipment Q100 TA Instruments are used in this study, This does not permit this testing, although the outcomes are consistent with earlier studies. Therefore, the data from our earlier research is utilized for pure ionic liquids. They used a second-order Eq. 9 to correlate the specific heat capacity to temperature like Eq. 7.

$$c_\rho(T) = \sum_{i=0}^{2} a_i T^l \qquad (9)$$

Where a_i : the regression parameter (using the least-square method)

δa_i is the standard uncertainty

While the fit was described using the coefficient of determination, R^2

Thermal applications of Ionanofluids
In the past few decades, there has been significant progress in the understanding and application of ionic liquid. In-depth research has been done on ionic liquids [107–117], which has shown them to be competitive replacements for a variety of industrial and chemical manufacturing applications. Their potential and success are influenced by phase equilibrium, thermophysical characteristics, and synthesis flexibility, among other factors. These fluids can be used in a number of applications due to their solvent properties, capacity to transmit or store heat, and surface characteristics [118]. Ionic liquids also have large volumetric heat capacities and great chemical and thermal stabilities.

AToMech1-2023 Supplement   Materials Research Forum LLC
Materials Research Proceedings 36 (2023) 16-37   https://doi.org/10.21741/9781644902790-3

Additionally, they have a wide range of viscosities, good solvent characteristics, and low vapor pressures. These are merely a few extra benefits.

- Enhanced thermal stability and heat transfer.
- Clog-free microchannel cooling.
- Downsized systems.
- Reduced pumping power.

Ionic liquid has come a long way in the last few decades, both in terms of understanding and practical use. Ionic liquids have been the focus of in-depth study [107–117] and have proven to be competitive substitutes for a range of industrial and chemical manufacturing applications. Phase equilibrium, thermophysical properties, and synthesis adaptability all contribute to their potential and success. Due to its solvent qualities, ability to transmit or store heat, and surface properties, these class of fluids can be used in a variety of applications [118]. Additionally, ionic liquids have excellent chemical and thermal stabilities as well as high volumetric heat capacities. Additionally, they exhibit good solvent properties, a wide variety of viscosities, and minimal vapor pressures. These are only a few additional advantages.

Because of their exceptional qualities, they have been widely investigated as molecular solvent replacements for liquid-phase operations. Because of their outstanding qualities and their current and potential applications in the chemical process industries, ionic liquids are of great interest to scientists and chemical corporations. It has been found that the values of their thermophysical properties greatly affect the design of physicochemical processing and reaction units by directly affecting the design parameters and operation of apparatus such heat exchangers, distillation columns, and reactors [119]. There are numerous recent studies [120-122] that give data on various thermophysical characteristics of various ionic liquids. These studies also look at measurement methods, measurement errors, and potential use of these fluids as heat transfer fluids. According to the results of these experiments, ionic liquids have enormous promise for a variety of applications, particularly as novel heat transfer fluids. However, for the optimal technical design of the green process, the characterization of the used ionic liquids, namely their thermodynamic, transport, and dielectric properties, is required.

The discovery that "bucky gels" can be produced by mixing carbon nanotubes and room-temperature ionic liquids opens up a brand-new field and has the potential to be applied in numerous engineering or chemical processes, such as the development of novel electronic devices, coating materials, and antistatic materials [52, 53]. The "bucky gels" are ionanofluids, which are emulsions or mixes of ionic liquids and nanomaterials, mostly nanocarbons, that are rich in CNTs (tubes, fullerenes, and spheres). There are many possible applications for ionic liquids that contain dispersed nanoparticles that have been functionalized in certain ways, such as functionalized single-walled carbon nanotubes (SWCNT), multiwalled carbon nanotubes (MWCNT), and fullerenes (C60, C80, etc.). By utilizing nanoparticles as heat transfer enhancers, it is possible to create ionanofluids, which are highly flexible and may be customized (target-oriented) in terms of molecular structure to gain the ideal attributes required to fulfill a certain task. The complicated interactions between ionic liquids and nanomaterials in the ensuing complex emulsions may be to blame for this. Ionic liquids, the building blocks of ionanofluids, are more versatile since they may be manufactured or created for particular characteristics and functions.

In contrast to their basic ionic liquid counterparts, MWCNT-containing ionanofluids exhibit higher thermal conductivity (from 2% to 35%) and specific heat capacity, according to recent study by Nieto de Castro and colleagues [121, 122]. These ionanofluids have excellent characteristics, such as high thermal conductivity, large volumetric heat capacity, and nonvolatility, which enable the development of novel heat transfer fluids. Ionanofluids can be used to make new pigments for solar collector paint coatings that have higher solar absorbance and thermal emissivity than base paint. Further than the investigations carried out by this group, there is no other ionanofluids study available in the literature.

Future scope of research studies

The following are the potential areas of future research involving ionanofluids:

- To completely characterize the effect of nanoparticles on base ionic liquids, empirical correlations, conventional numerical methods, and experimental observations must be used more frequently. Therefore, first-principle techniques such as density functional theory (DFT) or molecular dynamics (MD) simulation tools may be used to precisely assess the impact of nanoparticle size, shape, and concentration on the characteristics of ILs-based nanofluids.
- New IL-based nanofluids' thermo-physical properties and thermal performance can be characterized with artificial intelligence based on machine learning at a low computational cost. A deep learning or machine learning model can be trained and validated using the results of a molecular dynamics simulation.
- But given the available training data and computational resources, machine learning techniques must be carefully selected.
- To fully characterize the thermal performance of ILs-based nanofluids in full-scale solar thermal power systems, it is necessary to conduct experimental benchmarking. It is necessary to characterize the thermophysical characteristics and thermal performance of ILs-based nanofluids under suitable operating conditions, i.e., the high temperature of a solar thermal power plant.

Conclusion

The development of ionanofluid as a heat transfer fluid for thermal applications was summarized in this publication. It covers thermophysical parameters (thermal conductivity, viscosity, density, and heat capacity), the meticulous manufacture of ionic liquid, measurements process, theoretical and empirical correlation, heat transfer applications of ionanofluids, and potential areas for further research.

The following finding and suggestions were made as a result of the detailed examination of ILs-based nanofluids:

- The density of IL-based nanofluids has not been the subject of many studies. The density of ILs-based nanofluids increases in comparison to base ILs because much denser nanoparticles are added to the base fluid. Density decreases slightly as temperature rises.
- In their particular heat, IL-based nanofluids display scattered behavior with different nanoparticles. Nanofluids based on graphene and SWCNTs have a lower specific heat than the foundation ILs. High heat capacity is displayed by nanofluids made of multi-walled carbon nanotubes (MWCNTs). To better understand the behavior of specific heat detraction or enhancement, a thorough experimental research of the heat capacity of ILs-based nanofluids is required. Specific heat is one of the most significant thermophysical parameters for any heat storage medium.
- All of the studies revealed that the thermal conductivity of ILs-based nanofluids was higher than that of base ILs and that it rose with the concentration of nanoparticles. The increase in thermal conductivity was attributed to the interfacial layer of base ILs into the nanoparticles and the interaction between ions and nanoparticles.
- However, further investigation is required to completely comprehend the rise in thermal conductivity of ILs-based nanofluids.

Materials Research Forum LLC
https://doi.org/10.21741/9781644902790-3

References

[1] M.C. Roco, Nanoparticles and nanotechnology research. Journal of Nanoparticle Research, 1(1) (1999) 1.

[2] C.A. Nieto de Castro, S.M.S. Murshed, M.J.V. Lourenço, F.J.V. Santos, M.L.M. Lopes, J.M.P. França, Ionanofluids: new heat transfer fluids for green processes development, in M.A. Inamuddin (Ed.), Green Solvents I, Springer, Dordrecht, Netherlands (2012) 233–249. https://doi.org/10.1007/978-94-007-1712-18

[3] E. Ettefaghi, A. Rashidi, H. Ahmadi, S.S. Mohtasebi, M. Pourkhalil, Thermal and rheological properties of oil-based nanofluids from different carbon nanostructures, International Communications in Heat and Mass Transfer 48 (2013) 178–182. https://doi.org/10.1016/j.icheatmasstransfer.2013.08.004

[4] A.P.C. Ribeiro, M.J.V. Lourenço, C.A. Nieto de Castro, Thermal conductivity of ionanofluids, 17th Symposium on Thermophysical Properties, Boulder, USA, (2009).

[5] R.D. Rogers, K.R. Seddon, Ionic Liquids—Solvents of the Future? Science 302 (2003) 792–793.

[6] P. Kubisa, Application of ionic liquids as solvents for polymerization processes, Prog. Polym. Sci. 29 (2004) 3–12.

[7] J. F. Wishart, Energy applications of ionic liquids, Energy Environ. Sci. 2 (2009) 956 961.

[8] P. Singh, K. Kumari, A. Katyal, R. Kalra, R. Chandra, Copper nanoparticles in ionic liquid: An easy and efficient catalyst for selective carba-Michael addition reaction. Catal. Lett. 127 (2009) 119–125.

[9] R.G. Reddy, Novel applications of ionic liquids in materials processing. J. Phys. Conf. Ser. 165 (2009) 012076.

[10] A.E. Jiménez, M.D. Bermúdez, Ionic liquids as lubricants of titanium-steel contact. Tribol. Lett. 40 (2010) 237–246.

[11] A.A. Minea, S.M.S. Murshed, A review on development of ionic liquid based nanofluids and their heat transfer behavior, Renew. Sust. Energ. Rev. 91 (2018) 584-599. https://doi.org/10.1016/j.rser.2018.04.021

[12] A.P.C. Ribeiro, S.I.C. Vieira, J.M. França, C.S. Queirós, E. Langa, M.J.V. Lourenço, S.M.S. Murshed, C.A. Nieto de Castro, Thermal properties of ionic liquids and ionanofluids, in: A. Kokorin (Ed.), Ionic Liquids: Theory, Properties, New Approaches, InTech, Rijeka, Croatia (2011) 37–60. https://doi.org/10.5772/603

[13] W. Wang, Z. Wu, B. Li, B. Sundén, A review on molten-salt-based and ionic-liquid based nanofluids for medium-to-high temperature heat transfer, J. Therm. Anal. Calorim. 136 (2019) 1037–1051. https://doi.org/10.1007/s10973-018-7765-y

[14] S. Mallakpour, M. Dinari, Ionic liquids as green solvents: progress and prospects, in: A. Mohammad, Inamuddin (Eds.), Green Solvents II: Properties and Applications of Ionic Liquids, Springer, Dordrecht, the Netherlands, (2012) 1–32.

[15] A. Kokorin, Ionic Liquids: Applications and Perspectives, InTech, Rijeka, Croatia, (2011).

[16] A. Maia, Room temperature ionic liquids: a "green" alternative to conventional organic solvents? MROC 8 (2011) 178–185. https://doi.org/10.2174/157019311795177826

[17]   J.M. Dealy, J. Wang, Melt Rheology and its Applications in the Plastics Industry, 2nd ed. Springer, Dordrecht, the Netherlands, (2013).

[18]   P.C. Mishra, S. Mukherjee, S.K. Nayak, A. Panda, A brief review on viscosity ofnanofluids, Int Nano Lett 4 (2014) 109–120. https://doi.org/10.1007/s40089- 014- 0126-3

[19]   P. Kumar, K.M. Pandey, Effect on heat transfer characteristics of nanofluids flowing under laminar and turbulent flow regime – a review, IOP Conf. Ser.: Mater. Sci. Eng. 225 (2017), 012168. https://doi.org/10.1088/1757-899X/225/1/012168

[20]   S.M.S. Murshed, P. Estellé, A state of the art review on viscosity of nanofluids, Renew. Sust. Energ. Rev. 76 (2017) 1134–1152. https://doi.org/10.1016/j.rser. 2017.03.113

[21]   M.T. Jamal-Abad, M. Dehghan, S. Saedodin, M.S. Valipour, A. Zamzamian, An experimental investigation of rheological characteristics of non- Newtonian nanofluids, Journal of Heat and Mass Transfer Research (JHMTR) 1 (2014) 17–23. https://doi.org/10.22075/jhmtr.2014.150

[22]   T. Miri, Viscosity and oscillatory rheology, in: I.T. Norton, F. Spyropoulos, P. Cox (Eds.), Practical Food Rheology: An Interpretive Approach, Wiley-Blackwell, Oxford, UK (2011) 7–28. https://doi.org/10.1002/9781444391060.ch2

[23]   S.A. Khan, J.R. Royer, S.R. Raghana, Rheology: Tools and Methods, National Academic Press, Washington, USA, (1997).

[24]   T.F. Tadros, Rheology of Dispersions: Principles and Applications, Wiley, Weinheim, Germany, (2010). https://doi.org/10.1002/9783527631568

[25]   T.G. Mezger, The Rheology Handbook: For Users of Rotational and Oscillatory Rhe ometers, 4th ed. Vincentz Network, Hanover, Germany, (2014).

[26]   C.R. Jacobs, H. Huang, R.Y. Kwon, Introduction to Cell Mechanics and Mechanobiology, Garland Science, New York, USA, (2013).

[27]   V. Falguera, A. Ibarz, Juice Processing: Quality, Safety and Value-Added Opportuni ies, CRC Press, Boca Raton, USA, (2014).

[28]   D.R. Heldman, D.B. Lund, C. Sabliov, Handbook of Food Engineering, CRC Press, Boca Raton, USA, (2018).

[29]   T.F. Tadros, Nanodispersions, De Gruyter, Berlin, Germany, (2016).

[30]   H. Chen, Y. Ding, Heat transfer and rheological behaviour of nanofluids – a review, in L. Wang (Ed.), Advances in Transport Phenomena, Springer, Berlin, Germany (2009) 135–177. https://doi.org/10.1007/978-3-642-02690-4_3

[31]   A.K. Sharma, A.K. Tiwari, A.R. Dixit, Rheological behaviour of nanofluids: a review, Renew. Sust. Energ. Rev. 53 (2016) 779–791. https://doi.org/10.1016/j.rser.2015.09.033

[32]   J.P. Meyer, S.A. Adio, M. Sharifpur, P.N. Nwosu, The viscosity of nanofluids: a review of the theoretical, empirical, and numerical models, Heat Transfer Engineering 37 (2016) 387–421. https://doi.org/10.1080/01457632.2015.1057447

[33]   H. Babar, M.U. Sajid, H.M. Ali, Viscosity of hybrid nanofluids: a critical review, Therm. Sci. 23 (2019) 1713–1754.

ATOMech1-2023 Supplement   Materials Research Forum LLC
Materials Research Proceedings 36 (2023) 16-37   https://doi.org/10.21741/9781644902790-3

[34]   K. Bashirnezhad, S. Bazri, M.R. Safaei, M. Goodarzi, M. Dahari, O. Mahian, A.S. Dalkılıça, S. Wongwises, Viscosity of nanofluids: a review of recent experimental studies, International Communications in Heat and Mass Transfer 73 (2016) 114–123. https://doi.org/10.1016/j.icheatmasstransfer.2016.02.005

[35]   I.M. Mahbubul, R. Saidur, M.A. Amalina, Latest developments on the viscosity of nanofluids, Int. J. Heat Mass Transf. 55 (2012) 874–885. https://doi.org/10.1016/j.ijheatmasstransfer.2011.10.021

[36]   H.D. Koca, S. Doganay, A. Turgut, I.H. Tavman, R. Saidur, I.M. Mahbubul, Effect of particle size on the viscosity of nanofluids: a review, Renew. Sust. Energ. Rev. 82 (2018) 1664–1674. https://doi.org/10.1016/j.rser.2017.07.016

[37]   V.Y. Rudyak, Viscosity of nanofluids. Why it is not described by the classical theories, Advances in Nanoparticles 2 (2013) 720–726. https://doi.org/10.4236/anp.2013.23037

[38]   S.M.S. Murshed, P. Estellé, Rheological characteristics of nanofluids for advance heat transfer, in: A.A. Minea (Ed.), Advances in New Heat Transfer Fluids, CRC Press, Boca Raton, USA (2017) 227–266.

[39]   R. Vajjha, D. Das, Rheology and CFD Studies of Nanofluids: Thermophysical Properties and Correlations, LAP Lambert Academic Publishing, Riga, Latvia, (2016).

[40]   W.H. Azmi, K.V. Sharma, R. Mamat, G. Najafi, M.S. Mohamad, The enhancement of effective thermal conductivity and effective dynamic viscosity of nanofluids – a re view, Renew. Sust. Energ. Rev. 53 (2016) 1046–1058. https://doi.org/10.1016/j.rser.2015.09.081

[41]   L.S. Sundar, K.V. Sharma, M.T. Naik, M.K. Singh, Empirical and theoretical correlations on viscosity of nanofluids: a review, Renew. Sust. Energ. Rev. 25 (2013) 670–686, https://doi.org/10.1016/j.rser.2013.04.003

[42]   B. Abreu, A. Válega, B. Lamas, A. Fonseca, N. Martins, M. Oliveira, On the assessment of viscosity variability by nanofluid engineering: a review, J Nanofluids 5 (2016) 23 36. https://doi.org/10.1166/jon.2016.1189

[43]   A.A. Nadooshan, H. Eshgarf, M. Afrand, Evaluating the effects of different parame ters on rheological behavior of nanofluids: a comprehensive review, Powder Technol. 338 (2018) 342–353. https://doi.org/10.1016/j.powtec.2018.07.018

[44]   H. Khodadadi, S. Aghakhani, H. Majd, R. Kalbasi, S. Wongwises, M. Afrand, A com prehensive review on rheological behavior of mono and hybrid nanofluids: effec- tive parameters and predictive correlations, Int. J. Heat Mass Transf. 127 (2018) 997–1012. https://doi.org/10.1016/j.ijheatmasstransfer.2018.07.103

[45]   A. Shakeel, H. Mahmood, U. Farooq, Z. Ullah, S. Yasin, T. Iqbal, C. Chassagne, M. Moniruzzaman, Rheology of pure ionic liquids and their complex fluids: a review, ACS Sustain. Chem. Eng. 7 (2019) 13586–13626. https://doi.org/10.1021/ acssuschemeng.9b02232

[46]   K. Vignarooban, X. Xu, A. Arvay, K. Hsu, A.M. Kannan, Heat transfer fluids for concentrating solar power systems-A review. Appl. Energy, 146 (2015) 383–396.

[47]   R. Malviya, A. Agrawal, P.V. Baredar, A Comprehensive review of different heat transfer working fluids for solar thermal parabolic trough concentrator. Mater. Today Proc. 46(11) (2021) 5490–5500.

Materials Research Forum LLC
https://doi.org/10.21741/9781644902790-3

[48]   H.A. Zaharil, M. Hasanuzzaman, Modelling and performance analysis of parabolic trough solar concentrator for different heat transfer fluids under Malaysian condition, Renew. Energy. 149 (2020) 22–41.

[49]   T. Conroy, M.N. Collins, J. Fisher, R. Grimes, Thermohydraulic analysis of single phase heat transfer fluids in CSP solar receivers. Renew. Energy. 129 (2018) 150–167.

[50]   S.E. Trabelsi, L. Qoaider, A. Guizani, Investigation of using molten salt as heat transfer fluid for dry cooled solar parabolic trough power plants under desert conditions. Energy Convers. Manag. 15 (2018) 253–263.

[51]   N. Boerema, G. Morrison, R. Taylor, G. Rosengarten, Liquid sodium versus Hitec as a heat transfer fluid in solar thermal central receiver systems. Sol. Energy. 86 (2012) 2293–2305.

[52]   T. Fukushima, T. Aida, Ionic liquids for soft functional materials with carbon nanotubes. Chemistry–A European Journal, 13(18) (2007) 5048-5058.

[53]   K.R. Seddon, A taste of the future, Nature materials, 2(6) (2003) 363-365.

[54]   S. U. S. Choi, J. A. Eastman, Enhancing Thermal Conductivity of Fluids with Nanoparticles (Argonne National Laboratory: Lemont, IL) (1995).

[55]   J.D. Holbrey, K.R. Seddon, R. Wareing, A simple colorimetric method for the quality control of 1-alkyl-3-methylimidazolium ionic liquid precursors, Green Chemistry, 3(1) (2001) 33-36.

[56]   T. Fukushima, A. Kosaka, Y. Ishimura, T. Yamamoto, T. Takigawa, N. Ishii, T. Aida, Molecular ordering of organic molten salts triggered by single-walled carbon nanotubes. Science, 300(5628) (2003) 2072-2074.

[57]   R. Taylor, S. Coulombe, T. Otanicar, P. Phelan, A. Gunawan, W. Lv, G. Rosengarten, R. Prasher, H. Tyagi, Small particles, big impacts: A review of the diverse applications of nanofluids, J. Appl. Phys. 113 (2013) 011301. https://doi.org/10.1063/1.4754271

[58]   Y. Feng, E.E. Michaelides, G. Żyła, D. Jing, X. Zhang, P.M. Norris, C.N. Markides, O. Mahian, A review of recent advances in thermophysical properties at the nanoscale: From solid state to colloids, Phys. Rep. 843 (2020) 1–81. https://doi.org/10.1016/j.physrep.2019.12.001

[59]   E.C. Okonkwo, I. Wole-Osho, I.W. Almanassra, Y.M. Abdullatif, T. Al-Ansari, An updated review of nanofluids in various heat transfer devices, J. Therm. Anal. Calorim. (2020) https://doi.org/10.1007/s10973-020-09760-2

[60]   O. Mahian, L. Kolsi, M. Amani, P. Estellé, G. Ahmadi, C. Kleinstreuer, J.S. Marshall, M. Siavashi, R.A. Taylor, H. Niazmand, S. Wongwises, T. Hayat, A. Kolanjiyil, A. Kasaeian, I. Pop, Recent advances in modeling and simulation of nanofluid flows-Part I: Fundamentals and theory, Phys. Rep. 790 (2019) 1–48. https://doi.org/10.1016/j.physrep.2018.11.004

[61]   M. Lomascolo, G. Colangelo, M. Milanese, A. de Risi, Review of heat transfer in nanofluids: Conductive, convective and radiative experimental results, Renew. Sustain. Energy.

[62]   W. Wang, Z. Wu, B. Li, B. Sundén, A review on molten-salt-based and ionic- liquid based nanofluids for medium-to-high temperature heat transfer, J. Therm. Anal. Calorim. 136 (2019) 1037–1051. https://doi.org/10.1007/s10973- 018- 7765- y

[63]   A. Sobti, R.K. Wanchoo, Thermal conductivity of nanofluids, Mater. Sci. Forum757 (2013) 111–137. https://doi.org/10.4028/www.scientific.net/MSF.757

[64]   D. Tomida, Thermal conductivity of ionic liquids. In (Ed.), Impact of Thermal Conductivity on Energy Technologies. IntechOpen. (2018). https://doi.org/10.5772/intechopen.76559

[65]   C.N. De Castro, S.S. Murshed, M.J.V. Lourenço, F.J.V. Santos, M.M. Lopes, J.M.P. França, Enhanced thermal conductivity and specific heat capacity of carbon nanotubes Ionanofluids, International Journal of Thermal Sciences, 62 (2012) 34-39.

[66]   L. Godson, B. Raja, D.M. Lal, S. Wongwises, Experimental investigation on the thermal conductivity and viscosity of silver-deionized water nanofluid, Exp. Heat Transfer 23 (2010) 317–332.

[67]   S.M.S. Murshed, K.C. Leong, C. Yang, Investigations of thermal conductivity and viscosity of nanofluids, Int. J. Therm. Sci. 47 (2008) 560–568.

[68]   A. Riahi, S. Khamlich, M. Balghouthi, T. Khamliche, T.B. Doyle, W. Dimassi, A. Guizani, M. Maaza, Study of thermal conductivity of synthesized Al_2O_3-water nanofluid by pulsed laser ablation in liquid, J. Molecular Liquids 304 (2020) 112694. https://doi.org/10.1016/j.molliq.2020.112694

[69]   S. Zhang, Z. Ge, X. Fan, H. Huang, X. Long, Prediction method of thermal conductivity of nanofluids based on radial basis function, J. Therm. Anal. Calorim. 141 (2020) 859 880. https://doi.org/10.1007/s10973-019-09067-x.

[70]   Z. Said, S.M.A. Rahman, M. El Haj Assad, A.H. Alami, Heat transfer enhance- ment and life cycle analysis of a Shell-and-Tube Heat Exchanger using stable CuO/water nanofluid, Sustain. Energy Technol. Assess. 31 (2019) 306–317. https://doi.org/10.1016/j.seta.2018.12.020

[71]   T.P. Teng, Y.H. Hung, T.C. Teng, H.E. Mo, H.G. Hsu, The effect of alumina/water nanofluid particle size on thermal conductivity, Appl. Therm. Eng. 30 (2010) 2213 2218. https://doi.org/10.1016/j.applthermaleng.2010. 05.036

[72]   M.P. Beck, Y. Yuan, P. Warrier, A.S. Teja, The thermal conductivity of alumina nanofluids in water, ethylene glycol, and ethylene glycol+ water mixtures. Journal of Nanoparticle research, 12(4) (2010) 1469-1477.

[73]   W. Yu, H. Xie, L. Chen, Y. Li, Enhancement of thermal conductivity of kerosene-based Fe_3O_4 nanofluids prepared via phase-transfer method. Colloids and surfaces A: Physicochemical and Engineering Aspects, 355(1-3) (2010) 109-113.

[74]   H. Li, L. Wang, Y. He, Y. Hu, J. Zhu, B. Jiang, Experimental investigation of thermal conductivity and viscosity of ethylene glycol based ZnO nanofluids, Appl. Therm. Eng. 88 (2015) 363–368. https://doi.org/10.1016/ j.applthermaleng.2014.10.071

[75]   Y. Ding, H. Alias, D. Wen, R.A. Williams, Heat transfer of aqueous suspensions of carbon nanotubes (CNT nanofluids). International Journal of Heat and Mass Transfer, 49(1-2) (2006) 240-35.

[76]   A. Amrollahi, A.M. Rashidi, M. Emami Meibodi, K. Kashefi, Conduction heat transfer characteristics and dispersion behaviour of carbon nanofluids as a function of different parameters, Journal of Experimental Nanoscience, 4(4) (2009) 347-363.

Materials Research Forum LLC
https://doi.org/10.21741/9781644902790-3

[77]   J.C. Maxwell, A treatise on electricity and magnetism (Vol.1). Clarendon press, (1873).

[78]   R.L. Hamilton, O.K. Crosser, Thermal conductivity of heterogeneous two-components system, Industrial & Engineering Chemistry Fundamentals, 1(3) (1962) 187-191.

[79]   S.M.S. Murshed, K.C. Leong, C. Yang, Thermophysical and electrokinetic properties of nanofluids-a critical review, Applied Thermal Engineering, 28(17-18) (2008) 2109 2125.

[80]   S.M.S. Murshed, K.C. Leong, C. Yang, Investigations of thermal conductivity and viscosity of nanofluids, International Journal of Thermal Sciences, 47(5) (2008) 560 568.

[81]   A. Wittmar, M. Gajda, D. Gautam, U. Dörfler, M. Winterer, M. Ulbricht, Influence of the cation alkyl chain length of imidazolium-based room temperature ionic liquids on the dispersibility of TiO_2 nanopowders, Journal of Nanoparticle Research, 15(3) (2013) 1-12.

[82]   J.G. Huddleston, A.E. Visser, W.M. Reichert, H.D. Willauer, G.A. Broker, R.D. Rogers, Characterization and comparison of hydrophilic and hydrophobic room emperature ionic liquids incorporating the imidazolium cation, Green chemistry, 3(4) (2001) 156-164.

[83]   K.R. Seddon, A. Stark, M.J. Torres, Viscosity and density of 1-alkyl-3 methylimidazolium ionic liquids (2002).

[84]   K. Dong, Q. Wang, X. Lu, Q. Zhou, S. Zhang, Structure, interaction and hydrogen bond, Structures and Interactions of Ionic Liquids, (2014) 1-38.

[85]   L.F. Zubeir, M.A. Rocha, N. Vergadou, W.M. Weggemans, L.D. Peristeras, P.S. Schulz, M.C. Kroon, Thermophysical properties of imidazolium tricyanomethanide ionic liquids: experiments and molecular simulation, Physical Chemistry Chemical Physics, 18(33) (2016) 23121-23138.

[86]   A. Wittmar, M. Ulbricht, Dispersions of various titania nanoparticles in two different ionic liquids, Ind. Eng. Chem. Res. 51 (2012) 8425–8433. https://doi.org/10.1021/ie203010x

[87]   N. Zhao, Evaluation of physical properties of ionic liquids (Doctoral dissertation, Queen's University Belfast) (2017).

[88]   J. Gao, P.M. Mwasame, N.J. Wagner, Thermal rheology and microstructure of shear thickening suspensions of silica nanoparticles dispersed in the ionic liquid [C_4mim][BF4], Journal of Rheology, 61(3) (2017) 525-535.

[89]   K. Ueno, K. Imaizumi, K. Hata, M. Watanabe, Colloidal interaction in ionic liquids: Effects of ionic structures and surface chemistry on rheology of silica colloidal dispersions, Langmuir, 25(2) (2009) 825-831.

[90]   H. Tokuda, S. Tsuzuki, M.A.B.H. Susan, K. Hayamizu, M. Watanabe, How ionic are room-temperature ionic liquids? An indicator of the physicochemical properties, The Journal of Physical Chemistry B, 110(39) (2006) 19593-19600.

[91]   J. Novak, M.M. Britton, Magnetic resonance imaging of the rheology of ionic liquid colloidal suspensions, Soft Matter, 9(9) (2013) 2730-2737

[92]   F.F. Zhang, F.F. Zheng, X.H. Wu, Y.L. Yin, G. Chen, Variations of thermophysical properties and heat transfer performance of nanoparticle-enhanced ionic liquids, Royal Society Open Science, 6(4), (2019) 182040.

[93]   V. Khare, M.Q. Pham, N. Kumari, H.S. Yoon, C.S. Kim, J.I. Park, S.H. Ahn, Graphene–ionic liquid based hybrid nanomaterials as novel lubricant for low friction and wear, ACS applied materials & interfaces, 5(10) (2013) 4063-4075.

[94]   J. Liu, F. Wang, L. Zhang, X. Fang, Z. Zhang, Thermodynamic properties and thermal stability of ionic liquid-based nanofluids containing graphene as advanced heat transfer fluids for medium-to-high-temperature applications, Renewable Energy, 63 (2014) 519-523.

[95]   W. Chen, C. Zou, X. Li, An investigation into the thermophysical and optical properties of SiC/ionic liquid nanofluid for direct absorption solar collector, Sol. Energy Mater. Sol. Cells 163 (2017) 157–163. https://doi.org/10.1016/j.solmat.2017.01.029

[96]   H. Xie, Z. Zhao, J. Zhao, H. Gao, Measurement of thermal conductivity, viscosity and density of ionic liquid [EMIM][DEP]-based nanofluids, Chinese Journal of Chemical Engineering, 24(3) (2016) 331-338.

[97]   K. Khanafer, K. Vafai, A critical synthesis of thermophysical characteristics of nanofluids, International Journal of Heat and Mass Transfer, 54(19-20) (2011) 4410 4428.

[98]   Y. Xuan, W. Roetzel, Conceptions for heat transfer correlation of nanofluids, Int. J. Heat Mass Transfer 43 (2000) 3701–3707.

[99]   M. Sharifpur, S. Yousefi, J.P. Meyer, A new model for density of nanofluids including nanolayer, Int. Commun. Heat Mass Transfer 78 (2016) 168–174. http://dx.doi.org/10.1016/j.icheatmasstransfer.2016.09.010

[100] R. Hentschke, On the specific heat capacity enhancement in nanofluids, Nanoscale research letters, 11(1) (2016) 1-11.

[101] A. Safaei, A.H. Nezhad, A. Rashidi, High temperature nanofluids based on therminol 66 for improving the heat exchangers power in gas refineries, Applied Thermal Engineering, 170 (2020) 114991.

[102] R.S. Vajjha, D.K. Das, B.M. Mahagaonkar, Density measurement of different nanofluids and their comparison with theory, Petroleum Science and Technology, 27(6) (2009) 612-624.

[103] K. Oster, C. Hardacre, J. Jacquemin, A.P.C. Ribeiro, A. Elsinawi, Understanding the heat capacity enhancement in ionic liquid-based nanofluids (ionanofluids), Journal of Molecular Liquids, 253 (2018) 326-339.

[104] I. Carrillo-Berdugo, R. Grau-Crespo, D. Zorrilla, J. Navas, Interfacial molecular layering enhances specific heat of nanofluids: Evidence from molecular dynamics, Journal of Molecular Liquids, 325 (2021) 115217.

[105] S.S. Murshed, C.N. de Castro, M.J.V. Lourenço, J. França, A.P.C. Ribeiro, S. I.C. Vieira, C.S. Queirós, Ionanofluids as novel fluids for advanced heat transfer applications, International Journal of Physical and Mathematical Sciences, 5(4), (2011) 579-582.

[106] D.R. Lide, CRC handbook of chemistry and physics (Vol. 85), CRC press (2004).

[107] J.M.P França, Thermal properties of ionanofluids (Doctoral dissertation, M. Sc. thesis, Faculdade de Ciências da Universidade de Lisboa, Portugal) (2010).

[108] K.N. Marsh, J.A. Boxall, R. Lichtenthaler, Room temperature ionic liquids and their mixtures—a review, Fluid Phase Equilibria, 219(1), (2004) 93-98.

AToMoch1-2023 Supplement   Materials Research Forum LLC
Materials Research Proceedings 36 (2023) 16-37   https://doi.org/10.21741/9781644902790-3

[109] S. Zhang, N. Sun, X. He, X. Lu, X. Zhang, Physical properties of ionic liquids: database and evaluation, Journal of physical and chemical reference data, 35(4) (2006) 1475 1517.

[110] S. Keskin, D. Kayrak-Talay, U. Akman, O. Hortaçsu, A review of ionic liquids towards supercritical fluid applications, The Journal of Supercritical Fluids, 43(1) (2007) 150 180

[111] M.J. Earle, K.R. Seddon, Ionic liquids: Green solvents for the future, Pure and applied chemistry, 72(7) (2000) 1391-1398.

[112] J. Holbrey, Heat capacities of common ionic liquids-potential applications as thermal fluids?. Chimica Oggi-Chemistry Today, 25 (2007) 24-26.

[113] P. Wasserscheid, T. Welton, Ionic liquids in synthesis, Weinheim: Wiley-Vch, 1 (2008) 145.

[114] M. Gaune-Escard, K.R. Seddon, Molten salts and ionic liquids: never the twain?. John Wiley & Sons (2012)

[115] C. Chiappe, D. Pieraccini, Ionic liquids: solvent properties and organic reactivity, Journal of Physical Organic Chemistry, 18(4) (2005) 275-297.

[116] A. Heintz, C. Wertz, Ionic liquids: A most promising research field in solution chemistry and thermodynamics, Pure and applied chemistry, 78(8) (2006) 1587-1593.

[117] C.A. Nieto de Castro, F. SANTOS, Measurement of ionic liquids properties. Are we doing it well?. Chimica oggi, 25(6) (2007) 20-23.

[118] J.M. França, C.A. Nieto de Castro, M.M. Lopes, V.M. Nunes, Influence of thermophysical properties of ionic liquids in chemical process design, Journal of Chemical & Engineering Data, 54(9), (2009) 2569-2575.

[119] C.A. Nieto de Castro, E. Langa, A.L. Morais, M.L.M. Lopes, M.J.V. Lourenço, F.J.V. Santos, M.S.C.S. Santos, J.S.L. Lopes, H.I.M. Veiga, M. Macatrão, J.M.S.S. Esperança, L.P.N. Rebelo, C.S. Marques, C.A.M. Afonso, Studies on the density, heat capacity, surface tension and infinite dilution diffusion with the ionic liquids [C$_4$mim][NTf$_2$], [C$_4$mim][dca],[C$_2$mim][EtOSO$_3$] and [aliquat] [dca], Fluid Phase Equilibr 294 (2010) 157–179

[120] A.P.C. Ribeiro, M.J.V. Lourenço, C.N. de Castro, Thermal conductivity of Ionanofluids. In Proceedings of the 17th Symposium on Thermophysical Properties, (2009) 21-29.

[121] C.A. Nieto de Castro, M.J. V. Lourenço, A.P.C. Ribeiro, E. Langa, S.I.C Vieira, P. Goodrich, C. Hardacre, Thermal properties of ionic liquids and ionanofluids of imidazolium and pyrrolidinium liquids, Journal of Chemical & Engineering Data, 55(2) (2010) 653-66.

[122] S.I.C Vieira, A. Ribeiro, M. Lourenço, C. Nieto de Castro, Paints with Ionanofluids as pigments for improvement of heat transfer on architectural and heat exchangers surfaces. In Proceedings of the 25th European symposium on applied thermodynamics, Saint Petersburg, Russia (2011).

AToMech1-2023 Supplement
Materials Research Proceedings 36 (2023) 38-46

Materials Research Forum LLC
https://doi.org/10.21741/9781644902790-4

Characterization of sustainable concrete made from wastewater bottle caps using a machine learning and RSM-CCD: towards performance and optimization

Nayeemuddin Mohammed[1,a] *, Andi Asiz[1,b], Hiren Mewada[2,c], Zahara Begum[3,d], Salma Begum[3,e], Shahana Khatun[3,f], and Tasneem Sultana[4,g]

[1]Civil Engineering Department, Prince Mohammed Bin Fahd University, P.O. Box 1664 Al Khobar 31952, Kingdom of Saudi Arabia

[2]Electrical Engineering Department, Prince Mohammad Bin Fahd University, P.O. Box 1664 Al Khobar 31952, Kingdom of Saudi Arabia

[3]Civil Engineering Department, Khaja Banda Nawaz University Bilalabad Colony, Kalaburagi, Karnataka 585104, India

[4]Artificial Intelligence and Machine Learning, Godutai Engineering College for Women, Sharnbasva University, S.B. Campus, Sharana Nagar, Kalaburagi, Karnataka 585105, India

[a]mnayeemuddin@pmu.edu.sa, [b]aasiz@pmu.edu.sa, [c]hmewada@pmu.edu.sa, [d]zahara.civil777@gmail.com, [e]erimransalma026@gmail.com, [f]shahana.civilkhatun@gmail.com, [g]tasneemsultana841@gmail.com

Keywords: Plastic Bottle Caps, Compressive Strength, Artificial Neural Network, Central Composite Design

Abstract. The properties of concrete, a widely used building material across the globe, have changed due to technological breakthroughs. Cement, sand, coarse aggregate, and water are the four components used to build concrete. Technological improvements increase human comfort, yet the environment is also harmed. Therefore, recycling and reuse are vital to environmental engineers because they help reduce the amount of plastic bottle garbage disposed of as solid waste. In this study, water-cement ratios of 0.5, 0.55, and 0.6 are used in lieu of concrete in various percentages, including 0, 6, and 12% of coarse aggregate replaced by water bottle caps, to analyze the behavior of concrete's compressive strength experimentally. Based on experimental results, models based on artificial neural networks—Levenberg Marquardt and Response Surface Methodology—Central Composite Design models were developed to forecast the final compressive strength of concrete made in part from plastic water bottles. The results demonstrate that for accurately predicting the properties of concrete, the ANN-LM model yields the best result, $R^2=0.98$, which is close to 1 and $R^2 = 0.85$ for RSM-CCD, respectively.

Introduction

Concrete is widely employed as a building material. It may be utilized for all kinds of concrete constructions because it is versatile. Concrete is a composite building material primarily made of cement, water, and fine particles. Coarse aggregate can be partially substituted with plastic water bottle lids. With a global yearly use of 20–30 billion tones, it ranks second in materials consumed after water. Undoubtedly, one of the issues that will affect society the most in the future and that we must confront and solve in every manner possible is the recycling of waste materials of all types. Both the building and plastic recycling sectors benefit from developing plastic structural materials that use recycled plastic. Plastics are widely used, contributing to an ever-increasing volume of solid waste. The latter is obtained significantly from plastic bottles used as drink and mineral water storage containers [1]. Disposing of used plastic bottles and metal caps from soft drink bottles is a severe problem for environmental engineers and involves either recycling or reusing.

P.E.T. and metal bottle caps are added to the concrete at levels of 0, 5, 1, and 1.5% by volume of the entire mixture. Compressive, split tensile, and flexural strength are also analyzed, and the results are compared to conventional concrete. The findings demonstrate that standard concrete gains strength compared to various bottle cap percentages [2]. The workability of discarded plastic bottle caps that have been crushed has decreased. When 5% of the coarse aggregate is replaced with waste-crushed plastic bottle caps, the compressive strength increases by about 6.7%. When 15% of the coarse aggregate is replaced with waste-crushed plastic bottle caps, the strength decreases by about 27.6% using the water-cement ratio of 4.2 and 0.40. To increase concrete's compressive strength and flexural strength, it is feasible to mix discarded, crushed plastic bottle tops with cement [3].

Green construction is a crucial strategy to preserve natural resources and lessen the number of materials in our landfills. It is a worldwide issue that is becoming more and more essential. Recently, scientists looked at the feasibility of utilizing used bottle caps to partially replace coarse aggregate in the manufacturing of concrete [4]. The microstructure and mechanical properties of UHPC mixtures, including L.W.A., at high temperatures. Using a UHPC matrix containing different dosages of L.W.A., the effect of increasing temperature and L.W.A. concentration on the compressive and flexural strengths of the UHPC mixture was evaluated. Additionally, temperature exposure was applied to the UHPC combination before and after scanning electronic microscopy (S.E.M.) [5].

The ideal proportion for attaining the highest strength values is 10% cement substitution with B.L.A. to provide maximum compressive strength and durability against sulphate attack and water absorption [6]. The use of R.S.M. and ANN models revealed that they were accurate and valuable forecasts of the stiffness modulus and rutting of asphalt concrete mixes incorporating IWPET substitute aggregates [7]. Comparing the two statistical models revealed that, for the two responses taken into consideration, the ANN model performed better than the R.S.M. model due to its higher determination coefficient (R^2) and lower prediction errors (RMSE and M.R.E.) than the R.S.M. model [8]. The two model's comparison revealed that R.S.M. performed better than ANN, with an R^2 coefficient of determination of 0.9959, closer to 1 [9]. All R.S.M. prediction outcomes are within a 10% margin of experimental outcomes. However, three of the ANN model anticipated results fell outside the 10% limit [10].

The 3D microstructural investigation [11] suggested that the interfacial adhesion between the aggregates and the cementitious materials decreased with higher partial replacement. This reduction in interfacial adhesion results in unsatisfactory hardened characteristics. R.S.M. and ANN were used to assess the effects of adding fine glass aggregate and condensed milk can fiber (Sn) on the compressive and splitting tensile strength at three different curing ages. The inclusion of fiber and the substitution of fine glass aggregate for natural sand enhance the compressive strength of concrete from 14.15 to 16.05 MPa after seven days, 25.09 to 26.60 MPa after 28 days, and 32.12 to 34.14 MPa after 56 days, respectively, by 1% and 20%. The compressive strength of concrete diminishes as the quantity of both factors rises more [12]. Manufactured PE aggregates were utilized to replace fine natural aggregates, whereas P.E.T. aggregates were used to replace natural coarse aggregates at eight different volumetric replacement levels: 5%, 10%, 15%, 20%, 25%, 30%, 35%, and 40% [13]. R.S.M. is beneficial in forecasting the fresh and hardened characteristics of steel fiber-reinforced concrete because of its high predictive efficiency. This eliminates the tediousness of repeated laboratory experiments and enables quick decision-making for building applications [14].

The numerical method would be a potential advancement in this field since it delivers realistic and accurate forecasts while resolving most problems connected with the time-consuming,

Materials Research Forum LLC
https://doi.org/10.21741/9781644902790-4

dangerous, and expensive experimental techniques required [15]. The current study uses plastic wastewater bottle caps in various percentages and water-cement ratios. After 28 days of curing, concrete samples are taken for a crushing test to determine the compressive strength of the concrete. Additionally, it is experimentally compared with traditional concrete. The model, prediction, optimization, and assessment were designed using the artificial neural network, Levenberg-Marquardt, response surface approach, and central composite design methodologies.

Methodology
Materials:
The cement utilized was ordinary Portland cement (O.P.C.), which an Eastern Cement Company makes. Initial and final settling durations are 120 and 280 mm, respectively, and cement has a standard consistency of 26% with a soundness test score of 0.9. Locally, the fine aggregate was sourced from a quarry site in the eastern province's Dammam Industrial District. Testing found that the fineness modulus was 2.93, and the specific gravity was 2.75. Therefore, the fine aggregate was considered based on a size of less than 4.75 mm. Basaltic rock that had been crushed was used as a coarse aggregate. The specific gravity was found to be 2.63, the aggregate size was 25 mm, and it will be more than 4.75 mm. Each of these elements and processes was tested following ASTM standards. The Prince Mohammad Bin Fahd University's civil engineering material laboratory is supplied with potable water that has undergone a thorough inspection to guarantee that it is free of bacteria and undesired material. All mixing and cures are done using this water. Waste plastic bottle caps provide an excellent recycling resource and may be purchased at the scrap market in the Dammam industrial district. By promoting the recycling of these caps, significant energy is saved, and the use of discarded plastic bottle caps in other sectors is brought into sharper focus.

The casting of concrete sample:
Concrete mix designs are carried out, and the material's physical characteristics are assessed in compliance with ASTM standards. The ratio of 1:1.7:3 was used for the concrete mix, and varied water-to-cement ratios of 0.50, 0.55, and 0.60 were used. As per ASTM standards, a mould with dimensions of 150x150x150 mm was employed. For simple concrete sample removal, oil was added to the surface of the moulds after they had been securely fastened with screws. Figure 1 shows the concrete mixing with partial replacement of concrete by plastic water bottle caps, casting, and crushing test of a concrete sample after 28 days of curing.

Figure 1 The mixing, casting and crushing test of concrete samples.

Crushing test of concrete sample specimen:
After being in the mould for 24 hours, the concrete sample was taken out and allowed to cure for 28 days. Following a 28-day curing period, the specimen sample was examined for compressive

strength in a compression testing machine with a 2000 kN capacity in line with ASTM regulations. Table 1 compares the proportion of bottle caps, water cement ratio, coarse aggregate, fine aggregate, cement, water, and compressive strength to standard concrete.

Table 1 Compressive strength after 28 days.

Serial No:	(%) of Bottle Caps	Comp. Strength in (N/mm²)
1	0	26
2	6	27.2
3	12	28.3
4	6	27.5
5	12	28.8
6	0	27.8
7	6	27
8	12	27.3

Preparation of data sets for ANN-LM and RSM-CCD model optimization:
This method used eight experimental datasets to create the ANN-LM and RSM-CCD prediction models. Table 2 addresses the bottle cap concrete mix percentage.

Table 2 Concrete datasets with a mix of bottle caps for the ANN-LM and RSM-CCD models.

Serial No:	(%) of Bottle Caps	W/C ratio	C.A. (Kg/m³)	F.A. (Kg/m³)	O.P.C. (Kg/m³)	Water (Kg/m³)	Compressive Strength (N/mm²)
1	0	0.5	1200	850	310	160	26.0
2	6	0.5	1200	850	250	160	27.2
3	12	0.5	1200	850	190	160	28.3
4	6	0.55	1200	850	250	175	27.5
5	12	0.55	1200	850	190	175	28.8
6	0	0.6	1200	850	310	190	27.8
7	6	0.6	1200	850	250	190	27
8	12	0.6	1200	850	190	190	27.3

Development of ANN-LM and RSM-CCD prediction models:
Synthetic neural network, the model was created using MATLAB R2020b's Levenberg Marquardt backpropagation and Response Surface Methodology-Central Composite Design to forecast the compressive strength of partial substitution of coarse aggregate by plastic bottle caps [16]. Ordinary Portland cement, fine aggregate, coarse aggregate, water cement ratio, water content, and plastic bottle caps were the six variables used to create the model [17]. Additionally, the compressive strength of the concrete after the sample cured for 28 days was employed as the dependent or output variable [18]. In modeling, the datasets were split into 70% training data, 15% testing data, and 15% validation data [19]. Multiple neurons in 2 to 3 layers were used for testing and validation.   For the prediction of the output factor, Design Expert version 11.1.2 (Stat-Ease) central composite design and face-centered approach were used [20]. The percentage of plastic bottle caps (%) was taken into account in this design model as a partial substitute for coarse aggregate codes (A) and (B) for the water-cement ratio, as indicated in table 3. The design used 2^n factorial runs (n=2), and the number of runs was calculated using the equation $2^n + 2n + n_c$ [21].

AToMech1-2023 Supplement
Materials Research Proceedings 36 (2023) 38-46

Materials Research Forum LLC
https://doi.org/10.21741/9781644902790-4

The model was designed with three different levels in mind: lower, middle, and higher [22]. The polynomial quadratic equation 1 establishes a relationship between input and output variables.

$$Y' = \beta_{o'} + \beta_{1'}A' + \beta_{2'}B' + \beta_{12'}AB' + \beta_{11'}A'^2 + \beta_{22'}B'^2 \tag{1}$$

Where Y = response variable, $\beta_{o'}$ is intercept, $\beta_{1'}$ and $\beta_{2'}$ were linear coefficients, β_{11} and β_{22} were the quadratic coefficients, and β_{12} was the interaction between the coefficients.

Table 3. Experimental variables factors and coded levels in RSM-CCD design.

Factor	Name	Units	Type	Min.	Max.	Coded Low	Coded High
A	(%) of Bottle Caps	%	Numeric	-1.0	1.0	-1 ↔ -1.00	+1 ↔ 1.00
B	W/C Ratio		Numeric	-1.0	1.0	-1 ↔ -1.00	+1 ↔ 1.00

Results and Discussion

A well-trained model using the MATLAB tool ANN-LM was generated using the input parameters as the percentage of plastic water bottle caps and water-cement ratio with the number of hidden neurons varying from 10 to 30 [23]. The coefficient of correlation R^2 of regression analysis was employed to measure the model's accuracy, which shows the relationship between the estimated and measured variables. Figure 2 shows the highest relationship between the experimental vs. predicted response and the data fall along line 45° where the network outputs equal the targets [24]. It is found that the coefficient of determination (R^2) for the response allows the accurate prediction with $R^2 = 0.998$, which is close to 1.

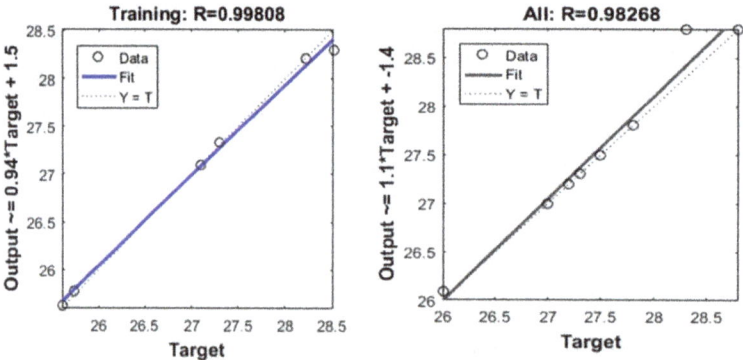

Figure 2 Experimental vs. Predicted values relationship for the response.

A central composite design and face-centered technique in response methodology were used to assess the experimental outcomes of a total of 8 runs with randomized type. The input factors' individual interface and quadratic impacts affect the prediction of the compressive strength of sustainable concrete, and an analysis of variance was conducted [25]. The R^2 predicted value was found to be 0.85. Figures 3(a) and (b) display the 3D plot between the input and output factor and prediction vs. actual experimental variables.

Analysis of variance (ANOVA) in design expert R.S.M. was used to look at the relationship between the inputs and outcome variables. The 5.36 model F-value suggests that the model terms are essential. A and B are important model terms in this instance. Model terms are insignificant if the value is higher than 0.10 [26].

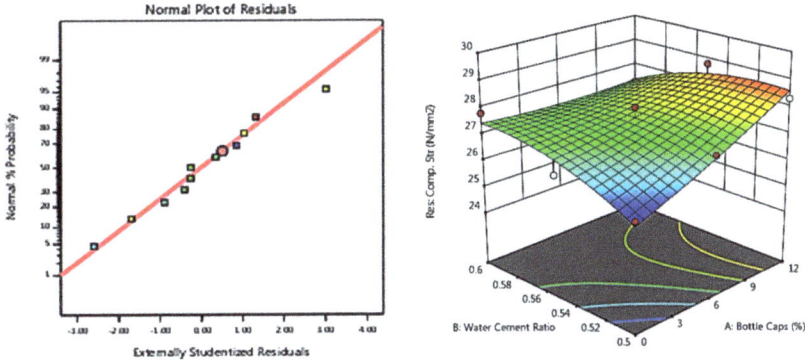

Figure 3 (left) A 3D plot and predicted vs. actual response (right) 3D surface plot

The lack of fit is implied to be insignificant compared to the pure mistake by the lack of fit F-value of 3.27. A significant lack of fit F-value has a noise probability of 24.28% occurring. The findings, which further corroborated the finding that allowed the null hypotheses of the two inputs to be rejected, reveal that the model's p-value is less than 0.05 [27]. Equations (2) and (3) show real and coded factors for response prediction. The coded factors may be used to predict responses based on independent factors using the coded factors. By comparing the factor coefficients, the quadratic equation in coded form proved extremely helpful in determining the relative importance of the elements [28]. The predictions may be made using the same equations in relation to the actual components.

Equation of compressive strength at 28 days in coded factors (N/mm^2) = $+27.61 + 0.6833(A) + 0.10(B) - 0.70(A.B.) + 0.1237(A^2) - 0.4263(B^2)$ (2)

Equation according to actual factors (N/mm^2) = $- 33.33 + 1.355(A) + 203.578(B) - 2.334(AB) + 0.00343(A^2) - 170.526(B^2)$ (3)

Where, A and B are input factors of the percentage of bottle caps and water cement ratio. Optimizing the mix proportion of bottle cap partial replacement of concrete was carried out using the optimization tool in the R.S.M. application [29]. In this tool setting, the maximum compressive strength of concrete on the 28th day predicted was found to be 28.6287 N/mm^2 at a percentage of bottle caps 12 % and water cement ratio of 0.5148 with the desirability of 0.961 1 out of 8 solutions as shown in figure 4 ramp diagram. Table 4 shows the compressive strength of concrete for the prediction by ANN-LM and RSM-CCD application.

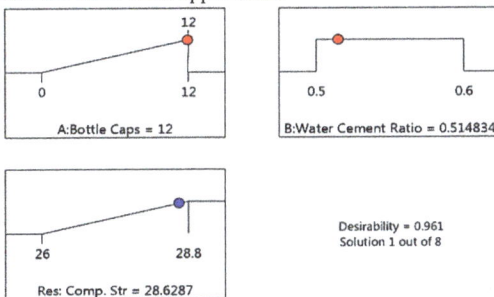

Figure 4 The desirability function of solution 1 out of 8

Table 4 Compressive strength of concrete by Expt., ANN-LM and RSM-CCD

Serial No:	Exp. Values (N/mm^2)	ANN-LM		RSM-CCD	
		Pred. Comp. Strength (N/mm^2)	(%) Error	Pred. Comp. Strength (N/mm^2)	(%) Error
1	26	26.09	-0.340	25.82	0.697
2	27.20	27.20	-0.100	27.08	0.443
3	28.30	28.80	-1.736	28.6	-1.049
4	27.5	27.51	0.036	27.61	-0.362
5	28.8	28.82	-0.035	28.42	1.407
6	27.80	27.82	-0.036	27.42	1.458
7	27	27.00	-0.917	27.28	-1.026
8	27.3	27.30	-0.365	27.39	-0.328

Conclusion

This method involved designing a concrete mix with various water-to-cement ratios and various percentages of plastic bottle tops to replace some of the coarse material. To ascertain the concrete sample's compressive strength, the experimental concrete mix was cast and examined after 28 days. For the prediction of concrete's compressive strength, software like RSM-CCD and ANN-LM were applied. The coefficient of determination of R^2 was determined to be 0.998 via ANN-LM. While the highest optimum value of concrete's compressive strength was determined to be 28.629 N/mm^2 at 12% of a 0.515 water-cement ratio, RSM-CCD was found to be 0.85. When predicting the compressive strength of concrete, ANN-LM and RSM-CCD are contrasted. Compared to RSM-CCD, ANN-LM modeling prediction was shown to have better values.

Conflict of Interest

The authors do not have any conflict of interest.

References

[1] D. Foti, Use of recycled waste pet bottles fibers for the reinforcement of concrete, Comp. Struct. 96 (12), 396-404. https://doi.org/10.1016/j.compstruct.2012.09.019

[2] S. Divyabharathi, S. Pavithran, Experimental study on mechanical properties of concrete by adding bottle caps and pet bottles in concrete, Int. Res. J. of Multi. Tech. 1 (2019) 490-495. https://doi.org/10.34256/irjmtcon70

[3] A.S. Awale, A.A. Hamane, Increase in strength of concrete by using waste plastic bottle caps as partial replacement of coarse aggregate, Int. Res. J. of Eng. And Tech. 7 (2020) 502-506.

[4] N. Khan, S.D. Agrawal, D.Y. Kshirsagar, Study of concrete by using waste plastic bottle caps as partial replacement of coarse aggregate, Int. Res. J. of Eng. And Tech. 4 (2017) 1699-1704.

[5] H. Alanazi, O. Elalaoui, M. Adamu, S.O. Alaswad, Y.E. Ibrahim, A.A. Abadel, A.F. Fuhaid, Mechanical and microstructural properties of ultra-high performance concrete with lightweight aggregates, Buildings. 12 (2022). https://doi.org/10.3390/buildings12111783

[6] G. Abebaw, B. Bewket, S. Getahun, Experimental investigation on effect on partial replacement of cement with bamboo leaf ash on concrete property, Adv. in Civ. Eng., (2021). https://doi.org/10.1155/2021/6468444

[7] M. Nayeemuddin, P. Palaniandy, Feroz S, Pollutants removal from saline water by solar photo catalysis: a review of experimental and theoretical approaches, Inter. J. of Environ. Anal. Chem. (2021) https://doi:10.1080/03067319.2021.1924160

[8] A. Usman, M.H. Sutanto, M. Napiah, S.E. Zoorob, N.S.A. Yaro, M.I, Khan, Comparison of performance properties and prediction of regular and gamma-irradiated granular waste polyethylene terephthalate modified asphalt mixtures, Polymers. 13 (2021). https:doi.org/10.3390/polym13162610

[9] M. Nayeemuddin, P. Palaniandy, Feroz S, Optimization of solar photocatalytic biodegradability of seawater using statistical modelling. J. the Indian Chem. Soc. 98(12), (2021). https://doi.org/10.1016/j.jics.2021.100240

[10] A.N. Rizalman, C.C. Lee, Comparison of artificial neural network (ANN) and response surface methodology (R.S.M.) in predicting the compressive strength of POFA concrete, model. And Simul. 4 (2020) 210-216.

[11] O.M. Ofuytan, O.B. Agbawhe, D.O. Omole, C.A. Igwegbe, J.O. Ighalo, R.S.M. and ANN modelling of the mechanical properties of self-compacting concrete with silica fume and plastic waste as partial constituent replacement, Clean. Mat. 4 (2022). https://doi.org/10.1016/j.clema.2022.100065

[12] S. Ray, M. Haque. T.Ahmed, T.T. Nahin, Comparison of artificial neural network (ANN) and response surface methodology (R.S.M.) in predicting the compressive and splitting tensile strength of concrete prepared with glass waste and tin (Sn) can fiber, J. of King Saud Uni. Eng. Sci. (2021). https://doi.org/10.1016/j.jksues.2021.03.006

[13] A. Shiuly, T. Hazra, D. Sau, D. Maji, Performance and optimization study of waste plastic aggregate based sustainable concrete – A machine learning approach, Clean. Waste Sys. 2 (2022). https://doi.org/10.1016/j.clwas.2022.100014

[14] T.F. Awolusi, O.L.Oke, O.O. Akinkurolere, A.O. Sojobi, Application of response surface methodology: predicting and optimizing the properties of concrete containing steel fibre extracted from waste tires with limestone powder as filler, Mat. 10 (2019). https://doi.org/10.1016/j.cscm.2018.e00212

[15] M.T.Mustafa, I. Hanafi, R. Mahmoud, B.A. Tayeh, Effect of partial replacement of sand by plastic waste on impact resistance of concrete: experiment and simulation, Structures. 20 (2019) 519-526. https://doi.org/10.1016/j.isrtuc.2019.06.008

[16] M. Nayeemuddin, Palaniandy P, F. Shaik, H. Mewada, D. Balakrishnan. Comparative studies of R.S.M. Box-Behnken and ANN-Anfis Fuzzy statistical analysis for seawater biodegradability using TiO_2 photo catalyst. Chem. (2023). https://doi.org/10.1016/j.chemosphere.2022.137665

[17] M. Nayeemuddin, Palaniandy P, F. Shaik. Solar photocatalytic biodegradability of saline water: Optimization using R.S.M. and ANN. A.I.P. Conf. Proc. 2463 (2022). https://doi.org/10.1063/5.0080297

[18] G. Nakkeeran and L.Krishnaraj, Prediction of cement mortar strength by replacement of hydrated lime using R.S.M. and ANN, Asian J. of Civ Eng. (2023). https://doi.org/10.1007/s42107-023-00577-6

[19] J. Sridhar, S. Balaji, D. Jegatheeswaran, P. Awoyera, Prediction of the mechanical properties of fibre-reinforced quarry dust concrete using response surface and artificial neural network techniques, Adv. in Civ. Eng, (2023). https://doi.org/10.1155/2023/8267639

[20] D.V. Dao, H.B. Ly, H.L.T Vu, T.T. Le, B.T. Pham, Investigation and optimization of the C-ANN structure in predicting the compressive strength of foamed concrete, Mat. 13 (2020). https://doi.org/10.3390/ma13051072

[21] B.D.O. Ayibiowu, O. Rufus, F. Kayode, Modelling the strength properties of concrete containing construction demolition waste using response surface methodology and artificial neural network, Eur. J. of App. Sci, 9 (2021). https://doi.org/10.14738/aivp.96.11464

[22] A. Nafees, M.F. Javed, S. Khan, K. Nazir, F. Farooq, F. Aslam, M.A. Musarat, N.I. Vatin, Predictive modeling of mechanical properties of silica fume-based green concrete using artificial intelligence approaches MLPNN, ANFIS and G.E.P., Mat. 14(2021). https://doi.org/10.3390/ma14247531

[23] M. Nazerian, M. Kamyabb, M. Shamsianb, M. Dahmardehb, M. Kooshaa, Comparison of response surface methodology (R.S.M.) and artificial neural networks (ANN) towards efficient optimization of flexural properties of gypsum-bonded fiberboards, Cerne. 24 (2018). https://doi.org/10.1590/01047760201824012484

[24] Mhaya, M. Akram, A.H. Amer, Shahidan, Shahiron, Zuki, S.S. Mohd, Azmi, M.A. Mohammad, Ibrahim, M.H.Wan, Huseien, G. Fahim, Systematic evaluation of permeability of concrete incorporating coconut shell as replacement of fine aggregate, Mat. 15 (2022). https://doi.org/10.339/ma15227944

[25] A. H. Amer, B. S.Abu, A. Rayed, S.A.R. Mohd, A.S. Alqarni, I.M.H. Wan, S.Shahiron, M. Ibrahim, A.B. Salami, Machine learning and R.S.M. models for prediction of compressive strength of smart bio-concrete, Kor. Sci, 28 (2021) https://doi.org/10.12989/sss.2021.28.4.535

[26] A. Hammoudi, K. Moussaceb, C. Belebchouche, F. Dahmoune, Comparison of artificial neural network (ANN) and response surface methodology (R.S.M.) prediction in compressive strength of recycled concrete aggregares, Cont. and Build. Mat. 209(2019), 425-436. https://doi.org/10.1016/j.conbuildmat.2019.03.119

[27] M.M. Hameed, M.K.Alomar, W.J. Baniya, M.A. AlSaadi, Prediction of high-strength concrete: high-order methodology modeling approach, Eng. With Comp, 38(2022), 1655-1668. https://doi.org/10.1007/s00366-021-01284-z

[28] M.N. Amin, M.F. Javed, K.Khan, F. Shalabi, M.G. Qadir, Modeling compressive strength of eco-friendly volcanic ash mortar using artificial neural networking, Symmetry, 13(2021). https://doi.org/10.3390/sym13112009

[29] G. Nakkeeran and L. Krishnaraj, Prediction of cement mortar strength by replacement of hydrated lime using R.S.M. and ANN, Asian, J. of Civ. Eng, (2023). https://doi.org/10.1007/s4210-023-00577-6

AToMech1-2023 Supplement
Materials Research Proceedings 36 (2023) 47-55

Materials Research Forum LLC
https://doi.org/10.21741/9781644902790-5

Pool boiling heat transfer characteristics of using nanofluids

Lujain Abdullatif Alshuhail[1,a] *, Alanood Mahmoud Almoaikel[1,b],
Feroz Shaik[1,c] and L. Syam Sundar [1,d]

[1]Department of Mechanical Engineering, Prince Mohammad Bin Fahd University, Al Khobar,
Kingdom of Saudi Arabia

[a]Lujain.Alshuhail@gmail.com, [b]anoodmu15@gmail.com, [c]ferozs2005@gmail.com,
[d]sslingala@gmail.com

Keywords: Pool Boiling, Nanofluids, Heat Transfer Rates, Thermal Conductivity, Heat Transfer Coefficient

Abstract Every day, smaller, faster, and more potent modern technologies and systems are being created and put into use, which necessitates the advancement of the thermal fluids that are used in operation to increase the capacity for heat removal. Pool boiling is effectively used in many industrial applications such as refrigeration systems, power plants etc. Application of nanofluids in pool boiling enhances the thermal conductivity and heat transfer rates in the system. This paper highlights the pool boiling heat transfer using nanofluids and its characteristics.

Introduction

Modern heat transfer technologies demand high heat flux rates. Fluid conductivity plays a significant role for high heat flux rates. Conventional heat transfer fluids such as air, water, ethylene glycol have poor thermal properties that limit the heat transfer equipment performances. Various researches reported using nanofluids in heat transfer equipment [1-4] but very few studies reported on pool boiling heat transfer using nanofluids. Pool boiling is most effective heat transfer application for heating and cooling of systems such as refrigeration systems, power plants etc. The heat transfer coefficient and rate of heat transfer are very high thus making it a crucial component in the use of energy dissipation systems.

Various methods are employed to improve heat transfer efficiency. One technique is to add solid nanoparticles to the heat transfer fluids, also known as nanofluids, to increase their thermal conductivity. These fluids are essential for transporting a lot of heat during the nucleate pool boiling phase change process. Heat produced by large-scale functioning equipment has been released through the boiling process. When the surface temperature is raised well above liquid saturation temperature, pool boiling occurs on the hot surface immersed in a pool of liquid. The movement of the liquid is the only byproduct of the heat transfer process, with no significant external donation [5]. Physiothermal properties such surface tension, viscosity, enthalpy, specific heat, thermal conductivity as well as the structure of the surface including the roughness and homogeneity are directly related to how heat is transferred in boiling pools. It also depends on the hydrodynamic condition close to the heating surface, such as the dynamics of hot and dry areas and the frequency, diameter of bubble departures [6].

As the nanofluid boils, nanoparticles simultaneously precipitate on the heated surface. The thickness of nanoparticles layer that has been formed is very thin at lower concentrations, making heat transfer and the magnitude of the heat transfer coefficient unaffected. In this scenario, the layer is ineffective and thermal conductivity or other heat transmission processes rule within the nanofluid. On the other hand, with greater nanofluid concentrations, the layer becomes thicker and the heat transfer coefficient suffers as a result. This happens because there are fewer active nucleation sites available over the heating surface due to the layer dominating the heat transfer

AToMech1-2023 Supplement
Materials Research Proceedings 36 (2023) 47-55

Materials Research Forum LLC
https://doi.org/10.21741/9781644902790-5

process, regardless of the nanofluid's capacity for heat transfer. Therefore, with nanofluid concentration, the critical heat flux remains constant and the heat transfer coefficient declines [7]. A study was conducted on pool boiling heat transfer using hybrid nanofluids with 0.01-0.1% volume concentrations. It was observed that hybrid nanofluids have a higher critical heat flux than single type nanofluids. Compared to single type nanofluids, hybrid nanofluids exhibit comparable stability and have a superior thermal conductivity. In comparison to deionized water, the maximum thermal conductivity improvement was 15.7% at $\phi = 0.1\%$. At $\phi = 0.01\%$ of hybrid nanofluids and in comparison to deionized water at critical heat flux, the maximum enhancement in h_{nf} was 7.1%. Deposition of nanoparticles on heater surface, the h_{nf} declines with the increase of ϕ value [8]. This paper presents the application of nanofluids in pool boiling heat transfer characteristics.

Effect of nanofluids thermal conductivity in pool boiling
In order to optimize heat transfer performance for diverse uses pool boiling, thermal conductivity is a crucial component. The advantages of enhanced thermal conductivities of various nanofluids are higher cooling/heating rates, low power requirements for pumping, thinner and lighter cooling/heating systems, reduced inventory of heat transfer fluids, reduced friction coefficients, and enhanced wear resistance. This improved the potential for the applications of nanofluids as refrigerants, cutting and hydraulic fluids, lubricants and coolants.

Base-fluid and nanoparticle thermal conductivity have a substantial impact on the enhancement of the thermal conductivity of nanofluids [9]. With ethylene glycol nanofluids compared to base-fluids, 0.3% copper nanoparticles showed a 40% improvement [10]. With base fluids and 0.1% of copper nanoparticles, thermal conductivity was claimed to be improved by 23.8% [11]. Through observation, it has been discovered that the surface to volume ratio of nanoparticles was a key influence in the improvement of heat conductivity [12]. An enhancement in thermal conductivity has been detected of almost 150% for poly oil with 1% volume fraction MWCNT [13]. For the same oil, but with 0.35% MWCNT, a 200% gain in thermal conductivity has been observed through experimental studies [14]. Cu nanoparticles in water with a 0.3% concentration improved thermal conductivity by 70% [15]. 75% increase in thermal conductivity has been observed for ethylene glycol with 1.2% diamond nanoparticles [16]. Nevertheless, research on thermal conductivity has produced some typical results [17–20]. Temperature, volume fraction, size, and structure of nanoparticles as well as pH, surfactant addition, nanofluid stability, and other elements all affect thermal conductivity [21-24].

Effect of nanofluids viscosity in pool boiling
The larger concentration of nanoparticles rises the viscosity of water-based nanofluids. As a result, the pressure drops over the cooling channel rises. Nanoparticle-water suspensions develop more viscous as the particle concentration in the suspension rises. The volume percentages of carbon nanotube are only acceptable to be less than 0.2% in real systems since the viscosity rose so quickly with higher particle loading. The particle mass fraction can therefore not be increased indefinitely. Replacing conventional fluids with nanofluids in industrial heat exchangers where significant volumes of nanofluids are required and turbulent flow is frequently created seems unfavorable [25,26]. More research studies need to be done on the effect of nanofluids viscosity in pool boiling.

Effect of nanofluids density in pool boiling
A study was conducted to observe the behavior of the density of nanofluids in pool boiling. The study's results reveal that the density of nanofluids behaves differently, and the data support the notion that this behavior is influenced by temperature. When raising the temperature, the density of the nanofluids is greater than that of its base fluid. The density falls as the volume concentration of nanoparticles rises [27]. Similar to viscosity, the density of a nano refrigerant rises as the volume

Materials Research Forum LLC
https://doi.org/10.21741/9781644902790-5

fraction rises and drops as the temperature rises. To achieve efficient energy performances, an optimal particle volume fraction should be determined, considering the thermal conductivity, viscosity, and density of the nano refrigerant (as thermal conductivity rises the heat transfer coefficients, whereas viscosity and density rises the pressure drop and pumping power) [28].

Effect of nanofluids specific heat in pool boiling
According to the literature, nanofluids have a lower specific heat than base fluid. In comparison to base fluids, CuO/ethylene glycol nanofluids, SiO_2/ethylene glycol nanofluids, and Al_2O_3/ethylene glycol nanofluids all have lower specific heats. Diamond nanoparticles are sediment over time thereby lowering the specific heat of the nanofluids. Nanofluids using in pool boiling should have a greater specific heat value in order to extract more heat from the environment [29,30]. Further research studies are needed to confirm the effect of nanofluids specific heat in pool boiling.

Effect of surface roughness in pool boiling
To study the effect of surface roughness an experiment investigations were performed using Al_2O_3 nanoparticles deposition in pool boiling heat transfer. The effect of deposition on critical heat flux of R-123 was studied. When compared to the uncoated surface, it was found that the surface coated with nanoparticles increased the critical heat flow by 17% and barely changed the heat transfer coefficient [31]. In other experimental studies, the effect of surface roughness and surface material was examined using two different nanofluids in pool boiling heat transfer. The experiments were conducted at reduced pressures using cylindrical surfaces made of stainless steel, brass, and copper. It was found that at low heat fluxes, the rough surface outperformed the smooth surface in terms of boiling thermal performance. However, this tendency shifted at high heat flux. The surface material had a significant impact on the slope between the heat transfer coefficient and heat transfer, which was larger for copper and brass but lower for stainless steel [32].

Effect of surface tension in pool boiling
Increasing the effectiveness and dependability of pool boiling heat transfer through the use of nanoparticles is an innovative, creative approach that has gain a noticeable interest in the past few years. Many parameters have been studied and reviewed yet the surface tension of the nanoparticles is still a vague area. In 2008, an experimental study and numerically simulation have been done on the migration characteristics of nanoparticles in the pool boiling process of Nano refrigerant and Nano refrigerant–oil mixture. This experiment aimed to determine whether the original mass of the nanoparticles affected their migration. The findings suggested that nanoparticles can migrate from the liquid phase to the gas phase during the pool boiling process using either a single or multiple individual escaping mechanisms. However, the liquid phase surface tension prevents such escaping.Nanoparticles with sufficiently high velocities and bubbles with attached nanoparticles break through the liquid's surface tension and float away. The surface tension of Nano refrigerant–oil mixture is higher than Nano refrigerant and hence the nanoparticle migration was greater in the later one [33].

In 2015, another experiment been conducted on the Surface tension of lithium bromide (LiBr) aqueous solution/ammonia with additives and nanoparticles. In this experiment to confirm the measuring accuracy, the surface tension of a LiBr aqueous solution with 1-octanol was measured. The obtained data were then compared with those from earlier experiments. In the study, additional chemicals including cetyltrimethylammonium chloride (CTAC) and cetyltrimethylammonium bromide (CTAB) were used. The outcomes of the experiment clearly demonstrate that CTAC and CTAB can decrease the surface tension of the LiBr aqueous solution/ammonia. It was also discovered that nanoparticles are unable to significantly lower the surface tension of LiBr aqueous solution/ammonia. However, the surface tension of LiBr aqueous solution/ammonia can be significantly changed by the combined addition of additives and nanoparticles. In other words, additives are more crucial in lowering the surface tension of the LiBr aqueous solution/ammonia. However, nanoparticles might improve heat transmission in the process of pool boiling [34].

Experimental studies on nanofluids pool boiling
Pool boiling is a well spread topic and there are research studies since 1962 but the research studies on application of nanofluids in pool boiling started recently. In 2020, Experimental research was done on the pool boiling heat transfer performance of deionized water and deionized water with magnesium-oxide nanoparticles. The nanofluids were generated at different volume concentrations and the pool boing heat transfer performance was tested under various heat flux and at atmospheric pressure. Using ultra sonication procedures, the stability of the created nanofluids was examined and was found to be reasonably good at least for the duration of the experiment. The obtained results showed that for volume concentrations of 0.001, 0.004, and 0.007 Vol%, the pool boiling heat transfer coefficient enhancement ratio was improved. The maximal enhancement ratio was 1.22 for 0.004 vol%. This ratio declined at the values of 0.01 and 0.04 vol% [35].

The tests were run with various heat flux and nanofluid concentrations for approximately 12 hours of boiling time. With various nanofluid concentrations and low and medium heat flow, the surface temperatures remained largely constant. For low, medium, and high concentration nanofluids, the high heat flow studies showed constant, oscillatory & incremental, and abrupt temperature spikes. With increasing heat flow, nanofluid concentrations, and boiling times, it was shown that the rate of heat transfer decreased by up to 90% and that the deposition of nanoparticles increased by 50% to 300%. Qualitative examinations of the microscopic pictures demonstrated the evolution of nucleation sites and deposition patterns over a range of trends as shown in Fig.1 [36].

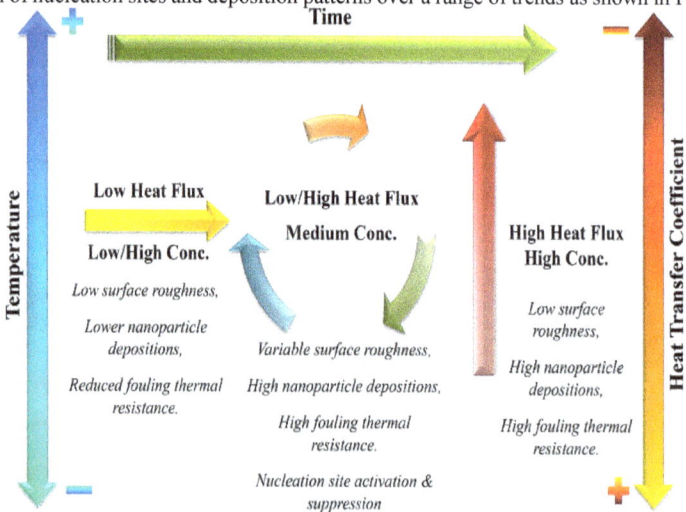

Figure 1. *Nucleation sites and deposition patterns evolve throughout a range of trends [36]*

In 2022, Experimental research was done on the pool boiling heat transfer of the Fe_3O_4/deionized water nanofluid while using mechanical vibration. The findings from the study of concentration also demonstrated that, at low concentrations, boiling heat transmission is increased while at high concentrations, it is decreased. As a result, at a concentration of 0.1 vol%, the boiling heat transfer coefficient was determined to be at its best for the nanofluid. When mechanical vibration is applied at all vibrational frequencies, it improves the boiling process' heat transmission. The maximum increase in the boiling heat transfer coefficient was found to be 87.3% when mechanical vibration was applied at the ideal concentration and vibrational frequency of 33 Hz [37].

Most recent experiment published in 2023, was experimental research done to examine and compare the pool boiling performance of a highly self-dispersion TiO_2 nanofluid with mass concentrations ranging from 0.0001 to 0.1%. Tests for repeatability, wettability, and macro/micro morphology were used to examine the impact of the boiling-induced deposition layer. The results showed that distilled water's critical heat flux (CHF) was only slightly increased by nanofluid, and that the CHF enhancement was caused by a super-hydrophilic micro-porous deposition layer that formed on the boiling surface of the nanofluid. Rising nanofluid concentration caused the deposition on the surface to occasionally come off, which complex and unpredictable boiling performance. Type and concentration of nanofluids had a minimal impact on CHF. The boiling curves of 0.0001% TiO_2 and Al_2O_3 nanofluids were comparable, although ZnO nanofluid had a lower heat transfer coefficient [38].

When nanoparticle and base fluid are combined, they are regarded as a single-phase mixture with stable qualities (mixed properties between the nanoparticle and base fluid properties). Given that the flow of a nanofluid may be thought of as a single-phase, incompressible flow, the simplest method for addressing the single-phase assumption is to use the governing equations for the flow of a pure fluid without taking the thermophysical characteristics of the nanofluid into account [39]. In investigating the "Linear stability theory of single phase nanofluids" it was concluded that the resulting eigen space of nano disturbances is constructed on the equivalent pure flow eigen space of perturbations, and that when the nanoparticles are added, the mean flow of nanofluids is somewhat modified [40].

Nanofluids/nanoparticles have been investigated heavily in the past two decades. It is known that nanoparticles increase the heat transfer rate. In recent study it has been shown that the boiling process increases vaporization velocity and has an impact on pressure and flow velocity as nanomaterial concentrations rise [41]. And recently based on experimental calculations it has been concluded that the heat transmission coefficient grows as the volumetric concentration of nanoparticles increases. When a result, as the heat flux increases, the boiling curve's slope decreases to lower superheat temperatures. The heat transfer coefficient increases by 49.35% by raising the volumetric concentration of nanoparticles to 0.1% when the heat flow is q = 341.8 kWm^{-2}. Additionally, the density of nucleation sites is inversely related to the quantity of nanoparticles. The diameter of the bubble leaving the zone with lower heat fluxes increases as the nanofluid concentration decreases. But as the nanofluid concentration drops at the high-flux zone, it gets smaller [42]. Therefore, we may infer that nanofluids can affect heat transfer boils in either a positive or negative way depending on their hydrophilicity and hydrophobicity.

To study the effect of hybrid nano fluids, an experiment has been done investigating TiO_2 and SiO_2 nanoparticles' effects on the nanofluid's heat transfer coefficient (HTC) during pool boiling. According to the findings, the HTC of TiO_2-SiO_2-water hybrid nanofluids is significantly higher than that of TiO_2 water and SiO_2 water single nanofluid systems. According to experimental data, the hybrid nanofluid is present at a concentration of 0.05% when the heat flux and HTC are at their greatest values [43].

There are many factors affecting pool boiling the most important ones are porosity, coating thickness, particle concentration, and surface roughness. An experimental study has been done in

2020 it studied the use of porous and the coating thickness, this experiment found that the use of porous heating surfaces enhances heat transfer efficiency and prevents temperature overshoot because of their linked porous structure, which increases wetted area and active nucleation site density. The outcomes also demonstrated that whereas low heat fluxes are best served by high thickness, high heat fluxes are best served by low thickness [44].

The surface roughness rises along with the nanofluid concentration, and the lower the nanofluid concentration, the smaller the contact angle of water with the coated surface. thus, increasing the diameter of the nanoparticles enhances the boiling heat transfer coefficient of nanofluids. The number of active bubble-generating sites is decreased when the size of the nanoparticles is decreased, filling smaller and more areas on the boiling surface. On the other hand, as nanoparticle diameter grows, more unoccupied places on the boiling surface of the copper block become accessible. Additionally, when nanoparticles build up, the boiling surface becomes rougher, creating potential new locations for the production of water vapor bubbles [45,46].

Conclusions

Pool boiling heat transfer is widely used in various industrial applications such as refrigeration systems, heating and cooling systems, power systems etc. Pool boiling with conventional fluids has limitations in heat transfer rates due to their Physiothermal properties. In the recent past, nanofluids with high thermal conductivity nanoparticles are widely applied for heat transfer applications. Researches are reported using nanofluids for pool boiling and it was observed the enhancement of heat transfer rates. However, there is a lot scope for further research studies on application of nanofluids in pool boiling, its heat transfer performance and effect of various Physiothermal properties.

References

[1] L.S. Sundar, F. Shaik, K.V. Sharma, V. Punnaiah, A.C.M. Sousa, The second law of thermodynamics analysis for longitudinal strip inserted nanodiamond-Fe3O4/water hybrid nanofluids. Int. J. Thermal Sciences. 181 (2022) 107721. https://doi.org/10.1016/j.ijthermalsci.2022.107721

[2] L.S. Sundar, M.K. Singh, A.C.M. Sousa, Enhanced heat transfer and friction factor of MWCNT-Fe3O4/water hybrid nanofluids, Int. Comm. Heat and Mass Transfer, 52 (2014) 73-83. https://doi.org/10.1016/j.icheatmasstransfer.2014.01.012

[3] L.S. Sundar, M.K. Singh, A.C.M. Sousa, Turbulent heat transfer and friction factor of nanodiamond-nickel hybrid nanofluids flow in a tube: An experimental study, Int. J. Heat and Mass Transfer, 117 (2018) 223-234. https://doi.org/10.1016/j.ijheatmasstransfer.2017.09.109

[4] L.S. Sundar, Feroz Shaik, Heat transfer and exergy efficiency analysis of 60% and 40% ethylene glycol mixture diamond nanofluids flow through a shell and helical coil heat exchanger, International Journal of Thermal Sciences, 184 (2023) 107901. https://doi.org/10.1016/j.ijthermalsci.2022.107901

[5] X.D. Fang, Y. Chen, H. Zhang, W. Chen, A. Dong, R. Wang, Heat transfer and critical heat flux of nanofluid boiling: A comprehensive review, Renewable and Sustainable Energy Reviews, 62 (2016) 924-940. https://doi.org/10.1016/j.rser.2016.05.047

[6] J. Buongiorno, L. Hu, I.C. Bang, Towards an Explanation of the Mechanism of Boiling Critical Heat Flux Enhancement in Nanofluids, In Proceedings of the ASME 2007 5th International Conference on Nanochannels, Microchannels, and Minichannels. ASME 5th International Conference on Nanochannels, Microchannels, and Minichannels, Puebla, Mexico, (2007) 989-995. https://doi.org/10.1115/ICNMM2007-30156

Materials Research Forum LLC
https://doi.org/10.21741/9781644902790-5

[7] B. Bharat, B.Divya, Nanofluids for heat and mass transfer, Academic Press, (2021).

[8] Y. Anil Reddy, S. Venkatachalapathy, Heat transfer enhancement studies in pool boiling using hybrid nanofluids, ThermochimicaActa, 672 (2019) 93-100. https://doi.org/10.1016/j.tca.2018.11.014

[9] Y.J. Hwang, Y.C. Ahn, H.S. Shin, C.G. Lee, G.T. Kim, H.S. Park et al., Investigation on characteristics of thermal conductivity enhancement of nanofluids, Current Applied Physics, 6(6) (2006) 1068-71. https://doi.org/10.1016/j.cap.2005.07.021

[10] J.A. Eastman, S.U.S. Choi, S. Li, W. Yu, L.J. Thompson, Anomalously increased effective thermal conductivities of ethylene glycol-based nanofluids containing copper nanoparticles, Applied Physics Letters, 78(6) (2001) 718-20. https://doi.org/10.1063/1.1341218

[11] M.S. Liu, M.C.C. Lin, I.T. Huang, C.C. Wang, Enhancement of thermal conductivity with CuO for Nanofluids, Chemical Engineering and Technology, 29(1) (2006) 72-7. https://doi.org/10.1002/ceat.200500184

[12] D.H. Yoo, K.S. Hong, H.S. Yang, Study of thermal conductivity of nanofluids for the application of heat transfer fluids, ThermochimicaActa, 455(1-2)(2007) 66-9. https://doi.org/10.1016/j.tca.2006.12.006

[13] S.U.S. Choi, Z.G. Zhang, W. Yu, F.E. Lockwood, E.A. Grulke,Anomalous thermal conductivity enhancement in nanotube suspensions, Applied Physics Letters, 79(14) (2001) 2252-4. https://doi.org/10.1063/1.1408272

[14] Y. Yang, Carbon nanofluids for lubricant application, University of Kentucky,(2006).

[15] S. Jana, A. Salehi-Khojin, W.H. Zhong, Enhancement of fluid thermal conductivity by the addition of single and hybrid nano-additives, ThermochimicaActa, 462(1-2) (2007) 45-55. https://doi.org/10.1016/j.tca.2007.06.009

[16] H.U. Kang, S.H. Kim, J.M. Oh, Estimation of thermal conductivity of nanofluid using experimental effective particle, Experimental Heat Transfer, 19(3) (2006) 181-91. https://doi.org/10.1080/08916150600619281

[17] X. Zhang, H. Gu, M. Fujii, Experimental study on the effective thermal conductivity and thermal diffusivity of nanofluids, International Journal of Thermophysics, 27(2) (2006) 569-80. https://doi.org/10.1007/s10765-006-0054-1

[18] X. Zhang, H. Gu, M. Fujii, Effective thermal conductivity and thermal diffusivity of nanofluids containing spherical and cylindrical nanoparticles, Journal pf Applied Physics, 100(4) (2006) 044325. https://doi.org/10.1063/1.2259789

[19] S. ZeinaliHeris, M.Nasr Esfahany, S.G. Etemad, Experimental investigation of convective heat transfer of Al2O3/water nanofluid in circular tube, International Journal of Heat and Fluid Flow, 28(2) (2007) 203-10. https://doi.org/10.1016/j.ijheatfluidflow.2006.05.001

[20] E.V. Timofeeva, A.N. Gavrilov, J.M.McCloskey, Y.V. Tolmachev, S. Sprunt, L.M. Lopatina, et al., Thermal conductivity and particle agglomeration in alumina nanofluids: experiment and theory, Physical Review E, 76(6) (2007) 16. https://doi.org/10.1103/PhysRevE.76.061203

[21] J.H. Lee, K.S. Hwang, S.P. Jang, B.H. Lee, J.H. Kim, S.U.S. Choi, et al., Effective viscosities and thermal conductivities of aqueous nanofluids containing low volume concentrations of Al2O3 nanoparticles, International Journal of Heat and Mass transfer, 51(11-12) (2008) 2651-6. https://doi.org/10.1016/j.ijheatmasstransfer.2007.10.026

[22] W. Yu, D.M. France, S.U.S. Choi, J.L. Routbort, Argonne National Laboratory review and assessment of nanofluid technology for transportation and other applications, Energy Systems Division, (2007). https://doi.org/10.2172/919327

[23] R.S. Vajjha, D.K. Das, Experimental determination of thermal conductivity of three nanofluids and development of new correlations, International Journal of Heat and Mass transfer, 52(21-22) (2009) 4675-82. https://doi.org/10.1016/j.ijheatmasstransfer.2009.06.027

[24] K.Y. Leong, R. Saidur, S.N. Kazi, M.A. Mamun, Performance investigation of an automotive car radiator operated with nanofluid based coolants (nanofluid as a coolant in a radiator), Applied Thermal Engineering (2010). https://doi.org/10.1016/j.applthermaleng.2010.07.019

[25] S. Wu, D. Zhu, X. Li, H. Li, J. Lei, Thermal energy storage behavior of Al2O3-H2O nanofluids, ThermochimicaActa, 483 (2009) 73-7. https://doi.org/10.1016/j.tca.2008.11.006

[26] K.L. Jin, K. Junemo, H. Hiki, T.K. Yong, The effects of nanoparticles on absorption heat and mass transfer performance in NH3/H2O binary nanofluids, International Journal of Refrigeration, 33 (2010) 269-75. https://doi.org/10.1016/j.ijrefrig.2009.10.004

[27] K. Habib, M. Ahmed, A.Q. Abdullah, O.A. Alawi, B. Bakthavatchalam, O.A. Hussein, Metallic Oxides for Innovative Refrigerant Thermo-Physical Properties: Mathematical Models, Tikrit Journal of Engineering Sciences, 29(1) (2022) 1-15. https://doi.org/10.25130/tjes.29.1.1

[28] I.M. Mahbubul, R. Saidur, M.A. Amalina,Thermal conductivity, viscosity and density of R141b refrigerant based nanofluid, Procedia Engineering, 56 (2013) 310-315. https://doi.org/10.1016/j.proeng.2013.03.124

[29] V. Trisaksri, S. Wongwises, Nucleate pool boiling heat transfer of TiO2-R141b nanofluids, Journal of Heat and Mass Transfer, 52(5-6) (2009) 1582-8. https://doi.org/10.1016/j.ijheatmasstransfer.2008.07.041

[30] K. Praveen, D.K. Namburu, K.M. Das, Tanguturi, S.V. Ravikanth, Numerical study of turbulent flow and heat transfer characteristics of nanofluids considering variable properties, International Journal of Thermal Sciences, 48 (2009) 290-302. https://doi.org/10.1016/j.ijthermalsci.2008.01.001

[31] Seok Bin Seo, In Cheol Bang, Effects of Al2O3 nanoparticles deposition on critical heat flux of R-123 in flow boiling heat transfer, Nuclear Engineering and Technology, 47(4) (2015) 398-406. https://doi.org/10.1016/j.net.2015.04.003

[32] J.M.S. Jabardo, An Overview of Surface Roughness Effects on Nucleate Boiling Heat Transfer, The Open Transport Phenomena Journal, 2 (2010) 24-34. https://doi.org/10.2174/1877729501002010024

[33] D. Ding, H. Peng, W. Jiang, Y. Gao, The migration characteristics of nanoparticles in the pool boiling process of nanorefrigerant and nanorefrigerant-oil mixture, International Journal of Refrigeration, 32(1) (2009) 114-123. https://doi.org/10.1016/j.ijrefrig.2008.08.007

[34] W.H. Cai, W.W. Kong, Y. Wang, M. S. Zhu, X.L. Wang, Surface tension of lithium bromide aqueous solution/ammonia with additives and nano-particles, Journal of Central South University, 22(5) (2015) 1979-1985. https://doi.org/10.1007/s11771-015-2718-0

[35] M.S. Kamel, F. Lezsovits, Experimental study on pool boiling heat transfer performance of magnesium oxide nanoparticles based water nanofluid, Pollack Periodica, 15(3) (2020) 101-112. https://doi.org/10.1556/606.2020.15.3.10

[36] A. Pare, S.K. Ghosh, The empirical characteristics on transient nature of al2o3-water nanofluid pool boiling, Applied Thermal Engineering, 199 (2021) 117617. https://doi.org/10.1016/j.applthermaleng.2021.117617

[37] M. Boroumand Ghahnaviyeh, A. Abdollahi, Experimental study of the effect of mechanical vibration on pool boiling heat transfer coefficient of Fe3O4/deionized water nanofluid, Journal of Thermal Analysis and Calorimetry, 147(24) (2022) 14343-14357. https://doi.org/10.1007/s10973-022-11591-2

[38] T. Wen, J. Luo, K. Jiao, L. Lu, Experimental study on the pool boiling performance of a highly self-dispersion TiO2 nanofluid on copper surface, International Journal of Thermal Sciences, 184 (2023) 107999. https://doi.org/10.1016/j.ijthermalsci.2022.107999

[39] S. Kakaç, A. Pramuanjaroenkij, Single-phase and two-phase treatments of convective heat transfer enhancement with nanofluids - a state-of-the-art review, International Journal of Thermal Sciences, 100 (2016) 75-97. https://doi.org/10.1016/j.ijthermalsci.2015.09.021

[40] M. Turkyilmazoglu, Single phase nanofluids in fluid mechanics and their hydrodynamic linear stability analysis, Computer Methods and Programs in Biomedicine, 187 (2020) 105171. https://doi.org/10.1016/j.cmpb.2019.105171

[41] H. ShakirMajdi, H.M. Abdul Hussein, L. JaaferHabeeb, D. Zivkovic, Pool boiling simulation of two nanofluids at multi concentrations in enclosure with different shapes of fins, Materials Today: Proceedings, 60, (2022) 2043-2063. https://doi.org/10.1016/j.matpr.2022.01.290

[42] S. Zaboli, H. Alimoradi, M. Shams, Numerical investigation on improvement in pool boiling heat transfer characteristics using different nanofluid concentrations, Journal of Thermal Analysis and Calorimetry, 147(19) (2022) 10659-10676. https://doi.org/10.1007/s10973-022-11272-0

[43] A. Mehralizadeh, S.R. Shabanian, G. Bakeri, Experimental and Modeling Study of heat transfer enhancement of TiO2/SiO2 hybrid nanofluids on modified surfaces in pool boiling process, The European Physical Journal Plus, 135(10) (2020). https://doi.org/10.1140/epjp/s13360-020-00809-7

[44] L.L. Manetti, A.S. Moita, R.R. de Souza, E.M. Cardoso, Effect of copper foam thickness on pool boiling heat transfer of HFE-7100, International Journal of Heat and Mass Transfer, 152 (2020) 119547. https://doi.org/10.1016/j.ijheatmasstransfer.2020.119547

[45] I.S. Kiyomura, L.L. Manetti, A.P. da Cunha, G. Ribatski, E.M. Cardoso, An analysis of the effects of nanoparticles deposition on characteristics of the heating surface and on pool boiling of water, International Journal of Heat and Mass Transfer, 106 (2017) 666-674. https://doi.org/10.1016/j.ijheatmasstransfer.2016.09.051

[46] A. Norouzipour, A. Abdollahi, M. Afrand, Experimental study of the optimum size of silica nanoparticles on the pool boiling heat transfer coefficient of silicon oxide/deionized water nanofluid, Powder Technology, 345 (2019 728-738. https://doi.org/10.1016/j.powtec.2019.01.034

AToMech1-2023 Supplement
Materials Research Proceedings 36 (2023) 56-62

Materials Research Forum LLC
https://doi.org/10.21741/9781644902790-6

Statistical analysis of fatigue behavior in additively manufactured steels

Ali Alhajeri[1], Mosa Almutahhar[1], Jafar Albinmousa[1,2], Usman Ali[1,2,3*]

[1]Department of Mechanical Engineering, King Fahd University of Petroleum & Minerals, Dhahran, 31261, Saudi Arabia

[2]Interdisciplinary Research Center on Advanced Materials, King Fahd University of Petroleum & Minerals, Dhahran, 31261, Saudi Arabia

[3]K.A. CARE Energy Research & Innovation Center at Dhahran, Saudi Arabia

usman.ali@kfupm.edu.sa

Keywords: Stainless Steel 316L, LBPF, Orientation, Condition, R-Value, Fatigue Behavior

Abstract. The effect of building orientations, sample conditions, and loading ratio (R-value) are important factors in terms of fatigue behavior. The aim of this paper is to investigate the factors that affect the fatigue behavior in additively manufactured laser powder-bed fusion (LPBF) 316L stainless steel. A statistical analysis was performed to point the significant and insignificant factors with different building orientations, samples conditions, and R-value. This statistical analysis provides the most significant factors to be considered for fatigue behavior of 316L stainless steel additive manufacturing.

1. Introduction

Conventional manufacturing processes have been extensively used for producing everyday industrial parts. Over the years, scientists and engineers have identified limitations in fabricating complex geometries using these techniques. In addition, conventional manufacturing processes result in wastage of material due to their subtractive nature [1]. Additive manufacturing (AM) provides a solution that can print complex geometries with little to no waste of material. Unlike the conventional manufacturing, AM simplifies the complexity of challenging geometries by manufacturing in a layer-by-layer fashion [2].

There are various commercially available AM technologies. However, laser powder-bed fusion (LPBF) is one of the most used AM process for industrial applications [2]. In this process, a laser is used as a source of thermal energy that fuses powder particles together to get the final shape in a layer-by-layer. Each layer is bonded to the next and previous layers to achieve the final part [2]. The process of LPBF involves a complex solidification and thermal cycle that can affect the development of microstructure. Since metal powders are the raw material used in LPBF, its performance can vary depending on the properties of the powder [3].

Besides cracks and surface deformation, other factors such as porosity, lack of fusion in powder particles, and stress risers can cause deterioration in the properties of LPBF. This can lead to an early catastrophic failure. In order to improve the performance of LPBF parts produced with various LPBF machines using similar process parameters, a comprehensive review of the available data is necessary. This process involves comparing the various studies that were conducted on the different test parameters, material sources and their sample conditions.

Several materials have been studied for analysis of mechanical properties for LPBF parts. Stainless steels have also been extensively studied due to their strength and applicability in producing functional components [4]. Stainless steel 316L is used in biocompatibility studies. These include internal fixation implants for hip joint surgeries [5], [6]. Besides biomedical

ATomMech1-2023 Supplement   Materials Research Forum LLC
Materials Research Proceedings 36 (2023) 56-62   https://doi.org/10.21741/9781644902790-6

applications, 316L is also widely used in various other industries such as aerospace, automotive, and the nuclear industry [4], [7]–[9]. Although 316L is widely used by various techniques, such as cutting, drawing, and stampeding, it is not easy to make final shape components due to its high work hardness, ductility, and low thermal conductivity. Due to these factors, it is often difficult to perform machining on 316L components. Using AM technology, which eliminates the need for a tool, it can be used to produce near-net-shape 316L components [8].

The objective of this study was to analyze the various factors that influence the fatigue performance of each factor. Through a multiple regression analysis, fatigue factors termed as significant factors were identified. Next, variance analysis (ANOVA) was performed to analyze the relationship of independent variables with the dependent variable. In this work, LPBF Stainless steel 316L fatigue data of un-notched samples was collected from literature and then used as input to the analysis software (Minitab ®). Results from the statistical analysis highlight the commonly used relationships already established in the statistical analysis along with an in-depth analysis of other factors.

2. Methodology

2.1 Factors of interest

There are many factors that affect the fatigue behaviors for different orientations and conditions as well as post-processing. The surface and part conditions are related to the surface roughness whether the samples are built to the net-shape or as a cylindrical or square rods and then machined. Also, polishing is an important factor can be applied to both machined and as-built samples [8]. Post-processing such as heat treatment (HT) such as annealing or hot isostatic pressure (HIP) applied to specimens can also be applied to samples which may affect their fatigue performance [10]. Different building orientations have also shown a pronounced effect on the fatigue behaviors of LPBF 316L parts [5]. Whether the samples were built vertically, horizontally or at any other intermediate angle can greatly affect the fatigue behavior.

Process parameters during fabrication of the samples is an important parameter as different authors use different machines and powder suppliers. In addition, different authors follow a slight variation of from the specific process parameters. Also, the material itself could cause a variation on the fatigue behaviors. This is due to the powder manufacturing process as each production company has different production system whether it's gas, water, plasma, atomization [11].

2.2 Statistical Analysis

ANOVA is a statistical technique that splits the aggregate variability in a data set into different parts, namely, the random and systematic factors. Although the former has a statistical influence, the latter does not. This allows the analysis of the relationship between independent and dependent variables and can be measured using the F-ratio. The F-ratio is also used to draw conclusions based on the assumptions of the random errors and variance. The null hypothesis in this analysis states that the results of the ANOVA's F-ratio test will be close to one if no real difference exists between the tested groups where the distribution of the F statistic follows the F-distribution [12]. The extracted data used in this work was sorted accordingly into a set of different independent groups which lead to a set of dependent fatigue performance responses. Then, statistical analysis was conducted via Minitab® statistical software. The corresponding results are presented below.

3. Results

The data collected for this study is based on 316L LPBF Stainless steel.

Table 1 shows the corresponding references with the conditions, orientations and R-values extracted from each paper. All data was analyzed ($\alpha = 0.05$) using statistical approaches as discussed in the previous section. Factors that affect the fatigue strength were analyzed as inputs with the maximum stress (σ_{max}) as the corresponding response.

*Table 1: Authors, conditions, orientations, and **R**-values for 316L analysis.*

Authors	Conditions	Orientation	R-Value
Shrestha et al. [7]	As built-HT Machined-Polish-HT		
Elangeswaran et al [8]	M As built Machined-HT As-built-HT		
Lai et al. [13]	As built-Polished Machined-Polished As built-Polished-HT Machined-Polished-HT	Z	-1
Afkhami et al. [9]	Machined As built HFMI		
Zhang et al. [14]	Machined	XY	
	Machined-Polish	Z	0.1

P-value analysis is one of the most used tools in statistical analysis of engineering analysis [12]. In addition to P-value analysis, Pareto charts can also be used to show the significance level of various factors. Table 1 shows the level of significantly for each factor.

Figure 1 shows the results of F-value where the dotted black horizontal line (at 1.97) highlights the significant factors. The results show that the number of cycles has the highest significance, then the conditions. Lastly, the R-value. Part orientation is not deemed as a significant factor in our analysis. However, there are various reports in literature where LPBF parts printed with different orientations show unequal responses [5].

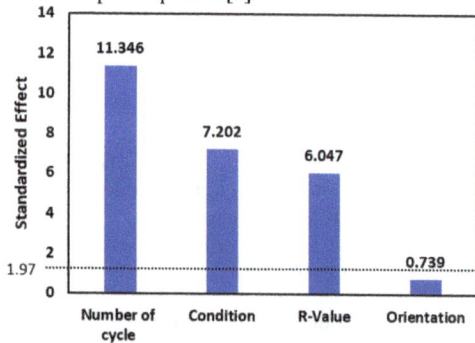

Figure 1: ANOVA table and Pareto Chart to identify the significant factors for 316L

Table 2: Analysis of variance results

Source	Seq SS	Cont.	Adj SS	Adj MS	F-Val	P-Val
Regression	6.1870	84.1%	6.18702	0.5624	79.57	0.00
Log Number of cycles	2.7062	36.8%	0.90999	0.9099	128.73	0.00
R-value	1.4221	19.3%	0.25847	0.2584	36.56	0.00
Orientation	0.0005	0.00%	0.00386	0.0038	0.55	0.46
Condition	2.0585	27.9%	2.05853	0.2573	36.40	0.00

Error	1.1663 15.8% 1.16637 0.0070		
Lack-of-Fit	1.1632 15.8% 1.16324 0.0070	2.26	0.49
Pure Error	0.0031 0.04% 0.00313 0.0031		

Figure 2 shows the investigation of why part orientation was not identified as a significant factor. Figure 2 shows the builds orientations from all publications with the same R-value [5], [8], [9], [13]. The results show that the statistical variance observed within the vertical samples contained all the results from the horizontal samples which deemed the orientation as insignificant. This is due to a lack of data for horizontal samples (only 1 study [9]).

To analyze the significant factors observed with the statistical analysis, R-value results from various authors were plotted as shown in Figure 3 [13], [14]. Experimental observations from various R-values (0.1, -1) show that the $R = 0.1$ partially overlaps with the deviation range of $R = -1$. This difference in the results between the two reported R-values results in a significant factor as shown in Figure 1.

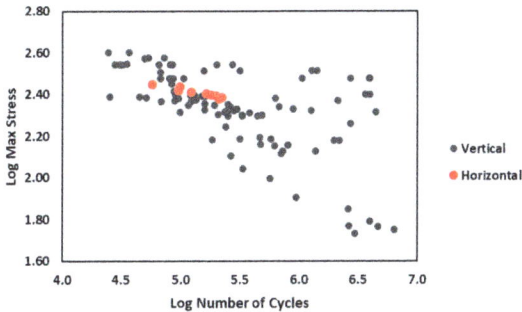

Figure 2: Fatigue data for different orientations 316L samples.

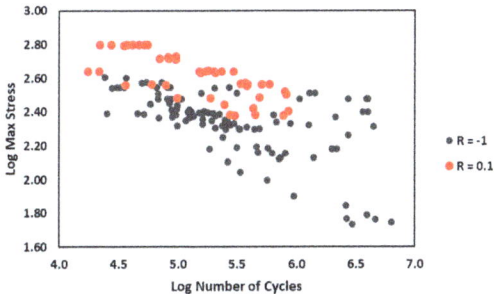

Figure 3: Fatigue data for different R-Values 316L samples.

It is important to analyze the statistical data to perform ANOVA. In this regard, several tools are used by researchers to identify if a certain set can be analyzed using statistical analysis. Normal probability plot is a graphical representation of the distribution of a given data set. It shows the likelihood that if or not the data set is distributed normally. Figure 4 shows that the fatigue data for 316L samples used in this study and shows a near normal distribution. It should be noted that a few points near 0.2 and -0.2 show minor deviations. This could be due to experimental error or anisotropic material behavior.

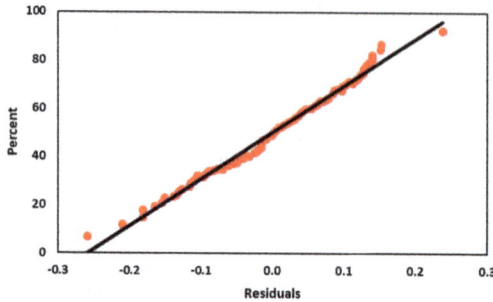

Figure 4: Normal probability plot for 316L samples.

A commonly used technique for performing a successful ANOVA is to create a scatter plot of the residuals and the fitted values. This type of plot is useful in detecting outliers, non-linearity, and unequal error variances. Figure 5 shows that the majority of the data correspond to a normal, equal error variance with few outliers.

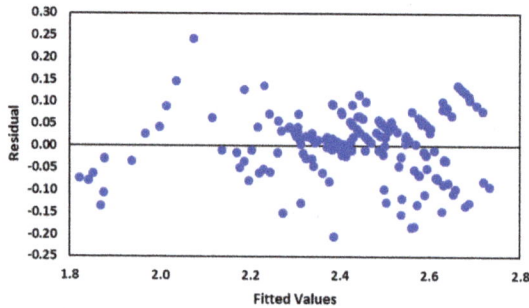

Figure 5: Residuals versus fitted values for 316L samples.

The use of an order plot (Figure 5) versus residual analysis is also useful in detecting the presence of non-independent error terms. It is used to identify the relationship between the various error terms in the sequence. Figure 6 shows that the fatigue data set has violated the independent error terms. Therefore, most experimental observations are independent from each other.

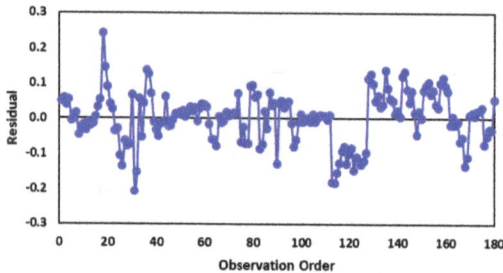

Figure 6: Residual versus observation order for 316L samples.

Finally, histograms are used to observe the dataset to identify anomalies in the recorded data. Histograms show the various data points grouped together into a logical range or bin. It can be used to compare the distribution of the given numerical data in intervals. It can also help an audience visualize and understand the various patterns and meanings of a data set. In addition, it can also be used to help the decision-making process of organizations. Figure 7 shows the maximum frequency for zero residuals along with a typical normal distribution of data as expected. The results from Figures 4-7 confirm that the data corresponds to a normal population and can be analyzed using a normal distribution. In addition, the results concluded from this analysis correspond to the common understanding in fatigue failure

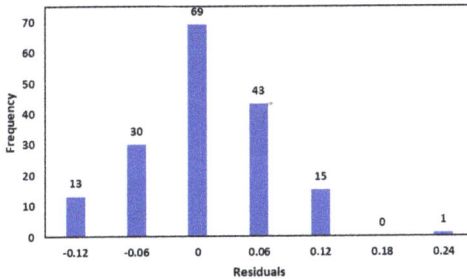

Figure 7: Histogram plot for 316L samples.

4. Conclusions

Fatigue failure data for Laser powder-bed fusion Stainless steel 316L from literature was collected and used in this work to conduct a statistical analysis on fatigue parameters. A few conclusions from this study are given below:

- Number of cycles, conditions, and R-values are identified as significant factors and therefore affect the fatigue strength significantly.
- Building orientation was not identified as a significant factor as the fatigue data of the horizontal build samples was limited and did not show major variation.
- Different R-values show partial significance when comparing $R = -1$ to $R = 0.1$.

Acknowledgements

The authors would like to acknowledge the help and support from Mechanical Engineering Department at KFUPM and would like to thank the financial support from King Abdullah City for Atomic and Renewable Energy (K.A.CARE). The authors would also like to acknowledge the help and support from the Rapid Prototyping and Reverse Engineering Lab at King Fahd University of Petroleum & Minerals.

References

[1] E. Atzeni and A. Salmi, "Economics of additive manufacturing for end-usable metal parts," *International Journal of Advanced Manufacturing Technology*, vol. 62, no. 9–12, pp. 1147–1155, Oct. 2012. https://doi.org/10.1007/s00170-011-3878-1

[2] F. Ahmed *et al.*, "Study of powder recycling and its effect on printed parts during laser powder-bed fusion of 17-4 PH stainless steel," *J Mater Process Technol*, vol. 278, Apr. 2020. https://doi.org/10.1016/j.jmatprotec.2019.116522

[3] J. Dawes, R. Bowerman, and R. Trepleton, "Introduction to the additive manufacturing powder metallurgy supply chain," *Johnson Matthey Technology Review*, vol. 59, no. 3. Johnson Matthey Public Limited Company, pp. 243–256, 2015. doi: 10.1595/205651315X688686

[4] G. S. Ponticelli, R. Panciroli, S. Venettacci, F. Tagliaferri, and S. Guarino, "Experimental investigation on the fatigue behavior of laser powder bed fused 316L stainless steel," *CIRP J Manuf Sci Technol*, vol. 38, pp. 787–800, Aug. 2022. https://doi.org/10.1016/j.cirpj.2022.07.007

[5] R. Shrestha, J. Simsiriwong, and N. Shamsaei, "Fatigue behavior of additive manufactured 316L stainless steel parts: Effects of layer orientation and surface roughness," *Addit Manuf*, vol. 28, pp. 23–38, Aug. 2019. https://doi.org/10.1016/j.addma.2019.04.011

[6] R. Shrestha, J. Simsiriwong, N. Shamsaei, S. M. Thompson, and L. Bian, "Effect of build orientation on the fatigue behavior of stainless steel 316l manufactured via a laser-powder bed fusion process," 2016. Accessed: Jan. 01, 2016. [Online]. Available: https://hdl.handle.net/2152/89615

[7] R. Shrestha, J. Simsiriwong, and N. Shamsaei, "Fatigue behavior of additive manufactured 316L stainless steel under axial versus rotating-bending loading: Synergistic effects of stress gradient, surface roughness, and volumetric defects," *Int J Fatigue*, vol. 144, Mar. 2021. https://doi.org/10.1016/j.ijfatigue.2020.106063

[8] C. Elangeswaran *et al.*, "Effect of post-treatments on the fatigue behaviour of 316L stainless steel manufactured by laser powder bed fusion." [Online]. Available: www.set.kuleuven.be/am/

[9] S. Afkhami, M. Dabiri, H. Piili, and T. Björk, "Effects of manufacturing parameters and mechanical post-processing on stainless steel 316L processed by laser powder bed fusion," *Materials Science and Engineering A*, vol. 802, Jan. 2021. https://doi.org/10.1016/j.msea.2020.140660

[10] F. Concli, L. Fraccaroli, F. Nalli, and L. Cortese, "High and low-cycle-fatigue properties of 17–4 PH manufactured via selective laser melting in as-built, machined and hipped conditions," *Progress in Additive Manufacturing*, vol. 7, no. 1, pp. 99–109, Feb. 2022. https://doi.org/10.1007/s40964-021-00217-y

[11] M. Jamshidinia, A. Sadek, W. Wang, and S. Kelly, "Additive Manufacturing of Steel Alloys Using Laser Powder-Bed Fusion," 2015. [Online]. Available: https://www.researchgate.net/publication/271831678

[12] Douglas C. Montgomery, *Design-and-Analysis-of-Experiments*, Ninth Edition. John Wiley & Sons, Inc., 2017. Accessed: Jan. 23, 2017. [Online]. Available: https://lccn.loc.gov/2017002355

[13] W. J. Lai, A. Ojha, Z. Li, C. Engler-Pinto, and X. Su, "Effect of residual stress on fatigue strength of 316L stainless steel produced by laser powder bed fusion process," *Progress in Additive Manufacturing*, vol. 6, no. 3, pp. 375–383, Aug. 2021. https://doi.org/10.1007/s40964-021-00164-8

[14] M. Zhang *et al.*, "Fatigue and fracture behaviour of laser powder bed fusion stainless steel 316L: Influence of processing parameters," *Materials Science and Engineering A*, vol. 703, pp. 251–261, Aug. 2017. https://doi.org/10.1016/j.msea.2017.07.071

Keyword Index

www.ingramcontent.com/pod-product-compliance
Lightning Source LLC
Chambersburg PA
CBHW071513210326
41597CB00018B/2736